循序渐进学 AI 系列丛书

从零开始学 TensorFlow 2.0

赵 铭 欧铁军 编著

电子工业出版社
Publishing House of Electronics Industry
北京·BEIJING

内 容 简 介

本书从 TensorFlow 2.0 的基础知识讲起，深入介绍 TensorFlow 2.0 的进阶实战，并配合项目实战案例，重点介绍使用 TensorFlow 2.0 的新特性进行机器学习的方法，使读者能够系统地学习机器学习的相关知识，并对 TensorFlow 2.0 的新特性有更深入的理解。

本书共 14 章，主要介绍机器学习、TensorFlow 2.0 基础、张量、数据层、CNN 等内容，中间还穿插了机器学习中常见的图形识别、文本处理和对抗训练等实例，以帮助读者理解 TensorFlow 2.0。本书着重介绍了在 TensorFlow 2.0 中使用 Keras 的方法，Keras 是 TensorFlow 2.0 中的重点概念，十分有必要对其进行学习。

本书内容通俗易懂、案例丰富、实用性强，特别适用于 TensorFlow 2.0 的入门者和进阶者，以及有志从事机器学习的爱好者，本书还适合用作相关机构的培训教材。

未经许可，不得以任何方式复制或抄袭本书之部分或全部内容。
版权所有，侵权必究。

图书在版编目（CIP）数据

从零开始学 TensorFlow 2.0/赵铭，欧铁军编著. —北京：电子工业出版社，2020.9
（循序渐进学 AI 系列丛书）
ISBN 978-7-121-39376-1

Ⅰ．①从… Ⅱ．①赵… ②欧… Ⅲ．①机器学习 Ⅳ．①TP181

中国版本图书馆 CIP 数据核字（2020）第 147444 号

责任编辑：李　冰
文字编辑：冯　琦
印　　刷：北京虎彩文化传播有限公司
装　　订：北京虎彩文化传播有限公司
出版发行：电子工业出版社
　　　　　北京市海淀区万寿路 173 信箱　邮编：100036
开　　本：787×1092　1/16　印张：16.75　字数：406 千字
版　　次：2020 年 9 月第 1 版
印　　次：2024 年 1 月第 3 次印刷
定　　价：79.00 元

凡所购买电子工业出版社图书有缺损问题，请向购买书店调换。若书店售缺，请与本社发行部联系，联系及邮购电话：（010）88254888，88258888。
质量投诉请发邮件至 zlts@phei.com.cn，盗版侵权举报请发邮件至 dbqq@phei.com.cn。
本书咨询联系方式：libing@phei.com.cn。

前言

这个技术有什么前途

目前，机器学习是 IT 领域最热门的话题之一，它能在看似无限的应用场景中发挥自身的作用，包括检测欺诈网站、自动驾驶及识别"金牌会员"身份以进行价格预测等。

通过机器学习，传统行业与互联网结合得更加紧密，机器学习能够帮助传统行业深度挖掘多年积累的数据，并根据各种行业场景制定模型。这些模型的合理应用，能够帮助各行业节省大量的人力和物力，为行业发展提供更多的数据支持。

随着各行业对机器学习的认识逐渐加深，通过选择合适的工具，从业人员可以简化建模过程，更专注地分析数据和设计算法。

TensorFlow 是机器学习领域的老牌开源软件，其适用性已经在机器学习领域得到了验证，其开放的学习社区和大量的学习资料能够为处于各阶段的学习者提供帮助。

笔者的使用体会

作为一直使用 TensorFlow 的互联网从业者，笔者在 TensorFlow 2.0 推出之际，首先使用其对原有项目进行了简单的升级。在该过程中，笔者发现，TensorFlow 2.0 根据 TensorFlow 社区众多开发者提出的意见在很多方面进行了优化，尤其是 Keras 的引入及其使用的加强，令笔者眼前一亮。

本书的特色

本书从基础的 TensorFlow 2.0 的安装、设置及应用开始介绍，并在介绍 TensorFlow 2.0 的新特性时使用了大量的实例，以帮助读者快速理解 TensorFlow 2.0 的特性。

TensorFlow 2.0 是一款机器学习工具，在介绍 TensorFlow 2.0 的同时，本书也穿插介绍了一些机器学习的基础知识，并以此为基础介绍了如何构建、训练和使用机器学习模型。

本书希望通过通俗易懂的示例来帮助读者理解深奥的算法知识，同时充分利用 TensorFlow 2.0 的新特性来保证读者能够学会使用机器学习工具，把读者从构建模型的繁杂工作中解放出来,使读者能更深刻地了解实际场景,分析场景中的逻辑并精炼算法,从而达到使用机器学习的目的。

本书的内容

第 1 章介绍了人工智能的概念和常用的机器学习软件。

第 2 章介绍了在 Linux 和 Windows 系统上安装与设置 TensorFlow 2.0 的方法,为后面使用 TensorFlow 2.0 做准备。

第 3 章介绍了 TensorFlow 2.0 的基础概念,如后面章节中用到的张量、数据集等。

第 4 章介绍了 TensorFlow 2.0 的应用:多层感知器。这是本书介绍的第一个 TensorFlow 2.0 的实际应用。

第 5 章深入介绍了卷积神经网络在 TensorFlow 2.0 中的应用。卷积神经网络是一种在深度学习中常用的网络模型结构。

第 6 章对 TensorFlow 2.0 的监督学习进行了介绍。

第 7 章对 TensorFlow 2.0 的新特性应用进行了介绍,介绍了如何使用 Keras 构建 TensorFlow 2.0 的网络模型并进行训练。

第 8 章针对典型的文本处理场景,介绍了如何使用 TensorFlow 2.0 对文本进行分类和处理。

第 9 章针对典型的图像处理场景,介绍了如何使用 TensorFlow 2.0 对图像进行分类和处理。

第 10 章通过实例介绍了决策树在 TensorFlow 2.0 下的使用。

第 11 章探讨了机器学习中常见的过拟合和欠拟合在 TensorFlow 2.0 下的优化方法。

第 12 章通过实例介绍了如何使用 TensorFlow 2.0 结构化数据。

第 13 章着重介绍了如何使用 TensorFlow 2.0 构建一个回归模型并进行训练。

作者介绍

赵铭:互联网 20 年从业者,目前就职于医疗大数据行业,从事数据仓库、数据分析和知识图谱等方面的研究。跟进了多个从 0 到 1 的项目,在项目调研、项目执行、项目推广和项目维护工作中均有不同程度的参与。曾在人人网担任基础架构工程师,在粉丝网担任 SRE 部门开发工程师。在多年的工作中,积累了一定的开发经验。

欧铁军:拥有 15 年软件和互联网工作背景。曾任 IBM 中国研究院研究员、高级软件工程师,成功完成多个 IBM 产品线的前沿研究工作,并在供应链、业务流程、智慧城市领域实施了多个大型项目;曾任国美库巴网 CTO,在国美收购库巴网一案中起到了关键作用。在之后的几年里,分别在 3 家创业公司担任 CTO,带领团队在云计算、O2O、C2B 领域完成了多次技术攻关。拥有多项计算机工程领域专利,发表了多篇学术论文。

本书的读者对象

- 机器学习的初学者
- 各数据公司的相关人员
- 各类培训班的学员
- 相关专业的大中专院校学生
- 需要工具书的学习者
- 其他对机器学习感兴趣的人

目录

第1章 人工智能的概念 1
   1.1 机器学习 1
   1.2 神经网络 3
   1.3 常用的深度学习框架 3

第2章 TensorFlow 初探 5
   2.1 在 Linux 系统中安装 TensorFlow 2.0 5
   2.2 在 Linux 系统中安装 TensorFlow 2.0 的 GPU 版本 5
   2.3 在 Windows 系统中安装 TensorFlow 2.0 6
   2.4 在 Windows 系统中安装 TensorFlow 2.0 的 GPU 版本 14

第3章 TensorFlow 的基础概念 17
   3.1 张量 17
   3.2 GPU 加速 19
   3.3 数据集 20
   3.4 自定义层 22
       3.4.1 网络层的常见操作 22
       3.4.2 自定义网络层 24
       3.4.3 网络层组合 25
       3.4.4 自动求导 26

第4章 TensorFlow 与多层感知器 30
   4.1 MLP 简介 30
   4.2 基础 MLP 网络 30
       4.2.1 回归分析 30
       4.2.2 分类任务 33
   4.3 基础模型 36
   4.4 权重初始化 39

4.5	激活函数	41
4.6	批标准化	44
4.7	dropout	46
4.8	模型集成	48
4.9	优化器	49

第 5 章 TensorFlow 与卷积神经网络 ... 52

5.1	基础卷积神经网络	52
5.2	卷积层的概念及示例	53
5.3	池化层的概念及示例	54
5.4	全连接层的概念及示例	55
5.5	模型的概念、配置及训练	57

第 6 章 TensorFlow 自编码器 ... 60

6.1	自编码器简介	60
6.2	卷积自编码器	64

第 7 章 TensorFlow 高级编程 ... 68

7.1	Keras 基础	68
	7.1.1 构造数据	68
	7.1.2 样本权重和类权重	70
	7.1.3 回调	72
7.2	函数式 API	76
	7.2.1 构建简单的网络	76
	7.2.2 构建多个模型	78
	7.2.3 两种典型的复杂网络	82
7.3	使用 Keras 自定义网络层和模型	86
	7.3.1 构建简单网络	86
	7.3.2 构建自定义模型	90
7.4	Keras 训练模型	94
	7.4.1 常见模型的训练流程	94
	7.4.2 自定义指标	96
	7.4.3 自定义训练和验证循环	100
7.5	Keras 模型的保存	106

第 8 章 TensorFlow 文本分类 ... 121

8.1	简单文本分类	121

8.2 卷积文本分类 ·131
8.3 RNN 文本分类 ·143

第9章 TensorFlow 图像处理 ·152
9.1 图像分类 ·152
9.2 图像识别 ·162
9.3 生成对抗网络 ·168

第10章 TensorFlow 决策树 ·180
10.1 Boosted Trees 简介 ·180
10.2 数据预测 ·180

第11章 TensorFlow 过拟合和欠拟合 ·197
11.1 过拟合和欠拟合的基本概念 ·197
11.2 过拟合和欠拟合 ·197
11.3 优化方法 ·208
    11.3.1 dropout 优化方案 ·208
    11.3.2 L2 正则化优化 ·212

第12章 TensorFlow 结构化数据 ·217
12.1 数字列 ·219
12.2 bucketized 列 ·220
12.3 类别列 ·222
12.4 嵌入列 ·223
12.5 哈希特征列 ·224
12.6 交叉功能列 ·226
12.7 结构化数据的使用 ·227

第13章 TensorFlow 回归 ·233
13.1 一元线性回归 ·233
13.2 多元线性回归 ·237
13.3 汽车油耗回归示例 ·241

第 1 章 人工智能的概念

目前,人工智能(Artificial Intelligence,AI)是一个比较热门的行业,从业人数众多且仍在快速增长。本章对人工智能的一些基本概念和流程进行介绍。

1.1 机器学习

传统的计算机程序需要先通过人工的方式给出限定条件,并做出相应的限制,程序按照限定条件对输入的数据集进行分析,并得到最终结果。而机器学习则是使用特定的训练集进行大规模数据计算以实现人工建立限制的过程,机器学习将数据(不限于文本数据、图形数据、音频数据和视频数据)提交给模型,模型根据算法对数据进行处理。

机器学习系统是通过训练得到的,不是显式的编制。机器学习过程是给机器学习一些与任务相关的例子,然后让机器通过例子推导并得出结果。深度学习是在机器学习基础上发展而来的一个重要方向,它通过特征建立相应的算法,从而达到让机器识别文字、数字和声音等数据的目的。传统程序、传统机器学习与深度学习之间的区别如图 1-1 所示。

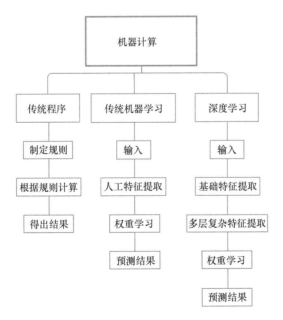

图 1-1 传统程序、传统机器学习与深度学习之间的区别

机器学习的本质是通过不断分析特征集建立函数算法的过程。可以将该过程大致分为以下 3 个步骤。

（1）选择一个合适的模型。针对不同的问题和任务的实际情况选取不同的模型，模型就是一组函数的集合。

（2）通过测试判断一个函数的好坏。

（3）找出合适的函数，常用的方法有梯度下降算法、最小二乘法等。

在选择了合适的函数后，要不断地在新样本上进行测试，函数只有在新样本上表现得好才算合适。

机器学习的核心是使用算法解析数据，使用数学方法建立函数，最终需要根据函数做出决定或进行预测。机器学习包括监督学习、无监督学习和强化学习。

这 3 种机器学习都有其特定的优点和缺点。

（1）监督学习需要有一组标记数据，使用特定模式来识别这种数据中每种标记的样本。监督学习主要包括分类、回归和排序。

① 在分类学习中，根据特定的模式将标记的数据划分为特定的类。

② 在回归学习中，根据特定的模式对已有数据进行处理以预测数据趋势。

③ 排序学习主要用于信息检索领域，需要按照一定的特征对特定数据集进行排序。

（2）无监督学习利用类别未知的训练样本解决数据识别的各种问题。需要注意的是，在无监督学习中，数据是无标签的。无监督学习包括聚类学习和降维学习。

① 聚类学习的实质是将数据集按照一定的特征进行分组。与分类学习不同，在聚类学习中不需要人为指定组的信息。

② 降维学习是通过在数据集中找到数据的共同点实现的。多数大数据可视化使用降维学习来探寻趋势和规则。

（3）强化学习使用机器的历史和经验来做决定。其不断地输入模糊数据，对机器学习函数进行强化，直至达到目的。强化学习是一个持续的过程。

常用场景分类如图 1-2 所示。

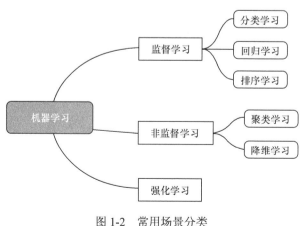

图 1-2　常用场景分类

1.2 神经网络

人工神经网络（Artificial Neural Network，ANN），简称神经网络或连接模型（Connection Model）。神经网络模仿了生物神经网络的特点，将生物神经网络数字化，使其更符合逻辑，从而模拟生物神经网络的"思考"过程，并将这个过程量化，保证了机器学习的准确率。神经网络的实际训练过程就是不断重复以上流程直至得到一个损失率较小的神经网络，然后使用训练过的神经网络模型对需要评测的数据进行处理。完整的神经网络训练流程如图1-3所示。

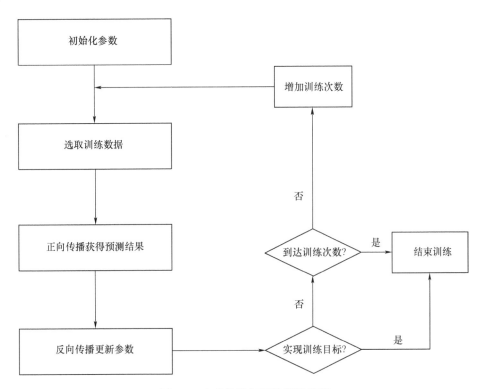

图1-3　完整的神经网络训练流程

1.3 常用的深度学习框架

选择一个合适的深度学习框架非常重要，每个深度学习框架都有其自身的特点，根据这些特点进行学习，能够起到事半功倍的效果。

最流行的深度学习框架有TensorFlow和PyTorch，具体情况如下。

（1）TensorFlow是使用人数最多、社区最庞大的框架，由Google公司开源，维护与更新比较频繁，并且有Python和C++的接口，教程也非常完善，是深度学习的主流框架之一。2019

年 3 月，TensorFlow 发布了 TensorFlow 2.0，与 TensorFlow 1.x 版本相比，其做了很多改变，比较重要的有：①重用 Keras 可以使模型的构建和运行更加简单；②强调各平台的通用性；③与 TensorFlow 1.x 相比，TensorFlow 2.0 提供了更方便的模型保存和调用方式。

（2）PyTorch 是从 Torch 框架演变来的深度学习框架，它使用 Python 在 Torch 框架上写了很多内容。不仅更加灵活，支持动态图，还提供了 Python 接口。PyTorch 由 Torch7 团队开发，能够实现强大的 GPU 加速，且支持动态神经网络。

第 2 章　TensorFlow 初探

TensorFlow 是一个基于数据流编程的符号数学系统，被广泛应用于各类机器学习（Machine Learning）算法的编程实现。

TensorFlow 拥有多层级结构，可部署于各类服务器、PC 终端和网页中，支持 GPU 和 TPU 高性能数值计算，广泛应用于谷歌的产品开发和各领域的科学研究。

本章介绍如何在 Linux 系统和 Windows 系统中安装 TensorFlow 2.0 的 CPU 版本和 GPU 版本。

2.1　在 Linux 系统中安装 TensorFlow 2.0

下面介绍如何在 Linux 系统中安装 TensorFlow 2.0。

（1）环境需求，本例中应用的系统环境如下。

```
Linux ubuntu 4.4.0-142-generic #168-Ubuntu SMP Wed Jan 16 21:00:45 UTC 2019 x86_64 x86_64 x86_64 GNU/Linux
```

（2）升级到 pip 最新版本（10.0.0 及以上版本），代码如下。

```
pip install pip -U
```

（3）直接安装 TensorFlow 2.0，代码如下。

```
pip install tensorflow==2.0.0
```

2.2　在 Linux 系统中安装 TensorFlow 2.0 的 GPU 版本

在 Linux 系统中安装 TensorFlow 2.0 的 GPU 版本需要 Nvidia 显卡的支持。因此，需要安装 cuda10。

（1）安装 Nvidia 驱动，需要安装 410.48 以上版本，禁止 Ubuntu 自带驱动，代码如下。

```
sudo vim /etc/modprobe.d/blacklist.conf        #编辑配置文件
```

（2）在打开的文件中添加如下代码。

```
blacklist nouveau
options nouveau modeset=0
```

（3）更新配置后验证，代码如下。

```
#更新配置
sudo update-initramfs -u
#重启
reboot
#检测是否禁止驱动，如果无输出，则禁止成功
lsmod | grep nouveau
```

(4) 安装 Nvidia 驱动,下载完成后执行如下命令。

```
sudo service lightdm stop                           #停止 lightdm 服务
cd install_package
sudo chmod 777 NVIDIA-Linux-x86_64-410.78.run       #开放相关文件执行权限
sudo ./NVIDIA-Linux-x86_64-410.78.run               #执行文件
```

(5) 重新启动图形界面,查看显卡驱动,命令如下。

```
#重启图形界面
sudo service lightdm start
#查看显卡驱动
nvidia-smi
```

(6) cuda 下载完成后进行安装,代码如下。

```
sudo chmod 777 cuda_9.0.176_384.81_linux.run        #开放 cuda 文件执行权限
sudo ./cuda_9.0.176_384.81_linux.ru                 #执行 cuda 安装文件
```

(7) 安装完成后配置到动态链接库,操作如下。

```
sudo gedit /etc/ld.so.conf.d/cuda.conf
```

(8) 添加如下语句。

```
/usr/local/cuda-10.0/lib64
```

(9) 保存退出后,执行如下命令。

```
sudo ldconfig
```

(10) 配置到环境变量,进行如下操作。

```
sudo gedit ~/.bashrc
```

(11) 在文件末尾添加如下路径。

```
export PATH=/usr/local/cuda-10.0/bin:$PATH
export LD_LIBRARY_PATH=/usr/local/cuda-10.0/lib64:$LD_LIBRARY_PATH
```

(12) 运行如下命令使配置生效。

```
source ~/.bashrc
```

(13) 下载 cuDNN 后进行安装,操作如下。

```
tar -zxvf cudnn-10.0-linux-x64-v7.5.1.10.tgz
cd cuda
sudo cp lib64/lib* /usr/local/cuda-10.0/lib64/
cd /usr/local/cuda-10.0/lib64/
sudo chmod +r libcudnn.so.7.5.1                     #查看.so 的版本,这里为 7.5.1 版本
sudo ln -sf libcudnn.so.7.5.1 libcudnn.so.7
sudo ln -sf libcudnn.so.7 libcudnn.so
```

(14) 安装 TensorFlow 2.0 的 GPU 版本,代码如下。

```
pip install tensorflow-gpu==2.0.0
```

2.3 在 Windows 系统中安装 TensorFlow 2.0

下面介绍如何在 Windows 系统中安装 TensorFlow 2.0,为了保证安装环境的一致性,采

用 Anaconda（开源的 Python 发行版本）配合安装。

（1）Anaconda 和 TensorFlow 2.0 的 Windows 版本都有最低安装要求，本例的安装环境详情如图 2-1 所示。

图 2-1 安装环境详情

（2）双击 Anaconda 安装文件，本例使用的 Anaconda 版本为 3-5.1.0 版本，弹出的 Anaconda 安装窗口如图 2-2 显示。

图 2-2 弹出的 Anaconda 安装窗口

(3)单击"运行"按钮,弹出窗口如图 2-3 所示。

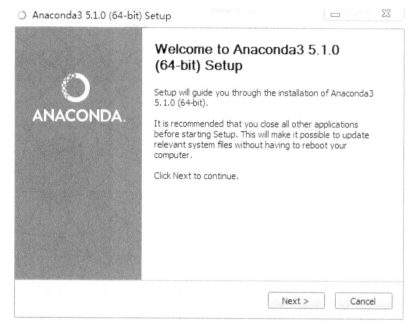

图 2-3　Anaconda 安装窗口 1

(4)单击"Next"按钮,弹出窗口如图 2-4 所示。

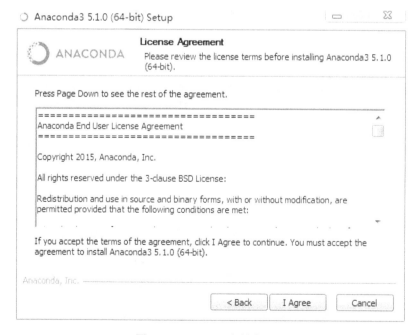

图 2-4　Anaconda 安装窗口 2

(5)单击"I Agree"按钮,弹出窗口如图 2-5 所示。

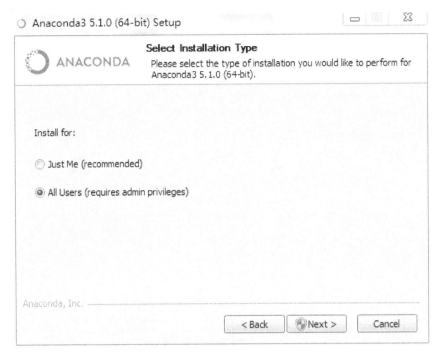

图 2-5　Anaconda 安装窗口 3

（6）单击"Next"按钮，弹出窗口如图 2-6 所示。

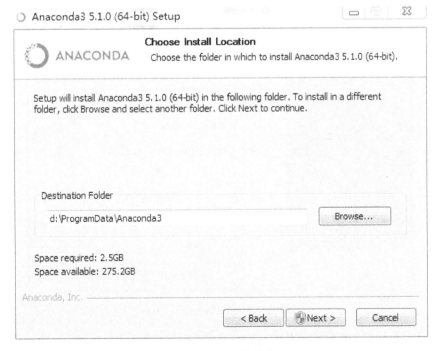

图 2-6　Anaconda 安装窗口 4

（7）选择安装目录后，单击"Next"按钮，弹出窗口如图 2-7 所示。

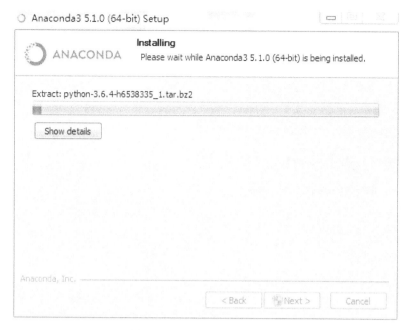

图 2-7　Anaconda 安装窗口 5

（8）等待安装完成后，单击"Next"按钮，弹出窗口如图 2-8 所示。

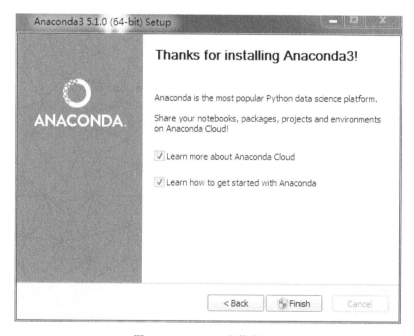

图 2-8　Anaconda 安装窗口 6

（9）单击"Finish"按钮完成安装。

（10）Anaconda 安装完成后需要对环境变量进行相应的测试，进入 Windows 中的命令模式，输入：conda --version，如果结果如图 2-9 所示则证明 Anaconda 已经安装完成。

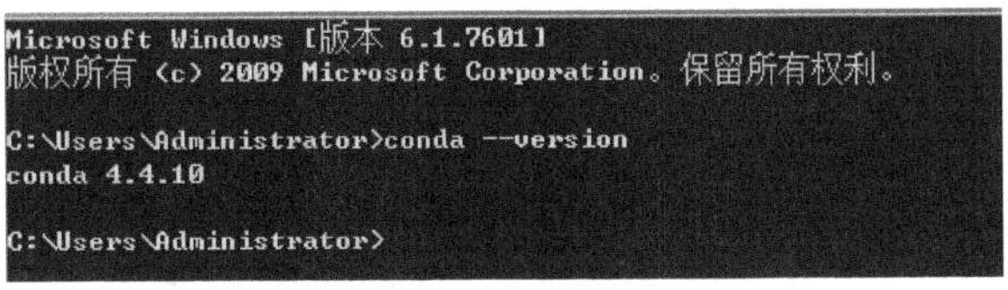

图 2-9　Anaconda 安装完成

（11）检测目前安装了哪些环境变量，在命令行中输入：conda info --envs，输出结果如图 2-10 所示。

图 2-10　Anaconda 环境变量

（12）由于 Anaconda 中安装了内置的 Python 版本解析器，这里基于 Python 进行介绍。在命令行中输入：conda search --full -name python，查看可用的 Python 版本，如图 2-11 所示。

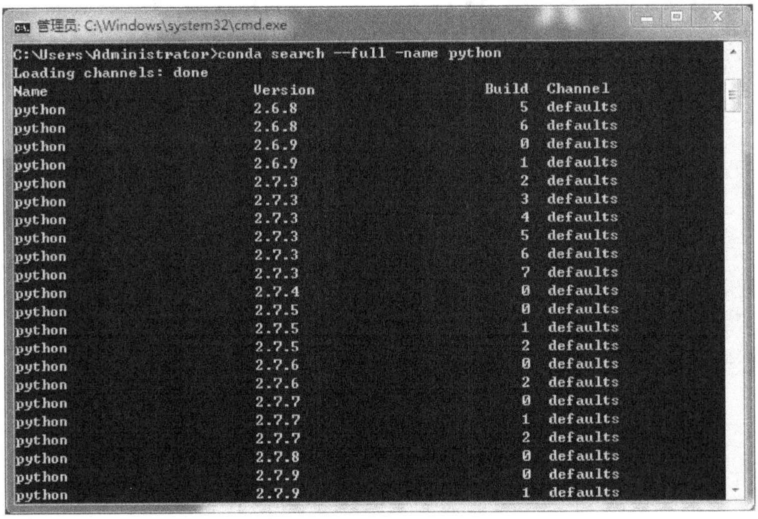

图 2-11　查看可用的 Python 版本

（13）安装 Python，这里安装 3.5 版本，在命令行中输入：conda create --name tensorflow python=3.5 进行安装，如图 2-12 所示。

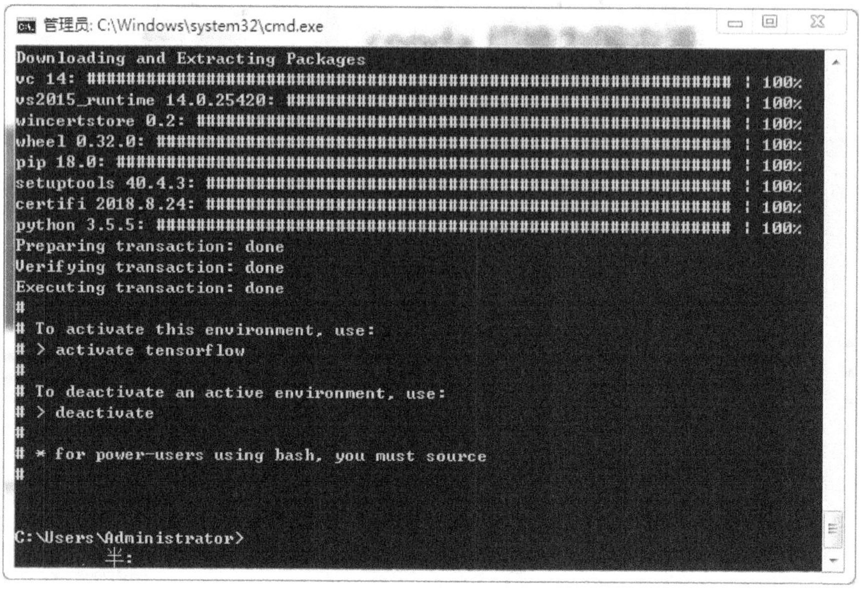

图 2-12　安装 Python

（14）激活 TensorFlow 环境，在命令行中输入：activate tensorflow，如图 2-13 所示。

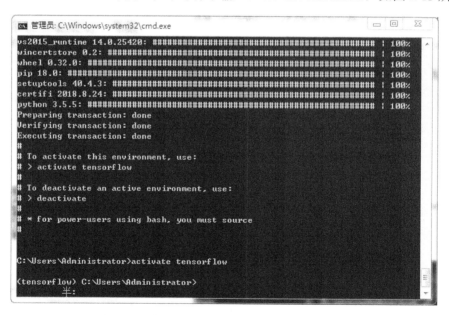

图 2-13　激活 TensorFlow 环境

（15）检测 TensorFlow 环境是否添加到 Anaconda 中，在命令行中输入：conda info --envs，如图 2-14 所示。

第 2 章　TensorFlow 初探

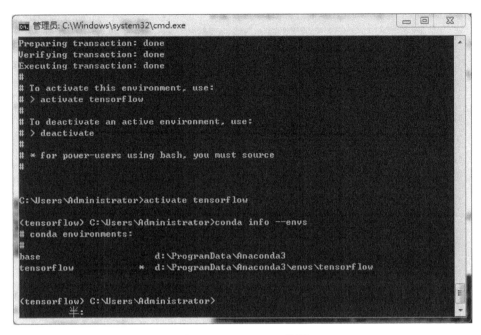

图 2-14　TensorFlow 检测

（16）检测当前环境中的 Python 版本，在命令行中输入：python --version，本例中的 Python 版本是 3.5.5，如图 2-15 所示。

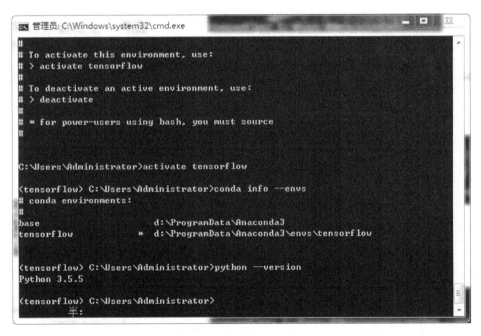

图 2-15　检测 Python 版本

（17）更新 TensorFlow 到最新版本，在命令行中输入：pip install --upgrade --default -timeout=100 --ignore -installed tensorflow 进行安装，如图 2-16 所示。

从零开始学 TensorFlow 2.0

图 2-16　更新安装 TensorFlow

2.4　在 Windows 系统中安装 TensorFlow 2.0 的 GPU 版本

在 2.3 节中安装了 TensorFlow 2.0 的默认版本，这个版本是基于 CPU 计算的，而 TensorFlow 在 GPU 下的运算效率会更高，本节举例说明如何在 Windows 下安装 TensorFlow 的 GPU 版本。

（1）安装 cuda 9.0、cuDNN 7.1，安装 GPU 版本与安装 CPU 版本类似，但是会多一步对 GPU 支持的安装。在安装前需要确认计算机拥有 Nvidia 的 GPU。

（2）在命令行中输入：conda create -n TF_2G python=3.5，构建 TensorFlow 2.0 的 GPU 环境，如图 2-17 所示。

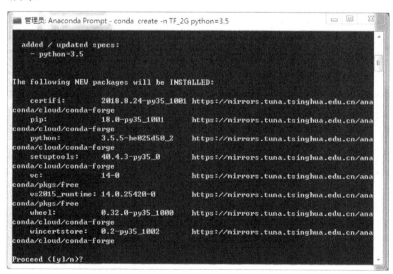

图 2-17　构建 TensorFlow 2.0 的 GPU 环境

第 2 章 TensorFlow 初探

（3）当弹出"Proceed ([y]/n)?"时输入"y"并按"回车"键，如图 2-18 所示。

图 2-18 新建 TensorFlow GPU 环境的操作

（4）完成后在命令行中输入：conda activate TF_2G，进入 TensorFlow 的 GPU 环境，如图 2-19 所示。

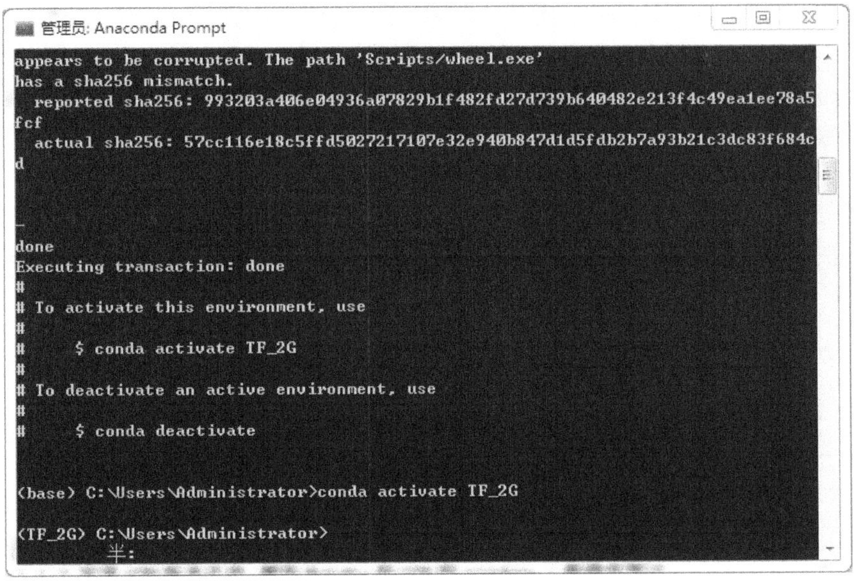

图 2-19 进入 TensorFlow 的 GPU 环境

（5）安装 GPU 版本支持，拥有 Nvidia GPU 的 Windows 一般都有默认驱动。因此，只需要安装 CUDA Toolkit 与 cuDNN 即可。在命令行中输入：conda install cudatoolkit=10.0 cudnn，如图 2-20 所示。

• 15 •

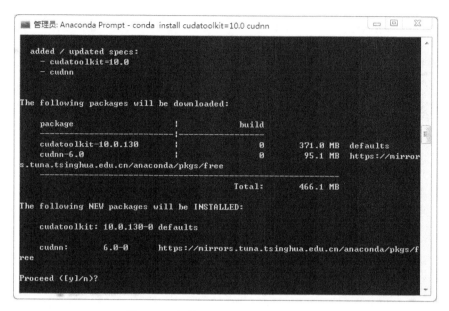

图 2-20 安装 CUDA Toolkit 和 cuDNN

注意：CUDA Toolkit 为 10.0.0 版本。

（6）安装 TensorFlow 2.0 的 GPU 版本，在命令行中输入：pip install tensorflow-gpu==2.0.0 -i https://pypi.tuna.tsinghua.edu.cn/simple，进行安装。

说明："-i"表示从国内清华源下载。

（7）测试 TensorFlow 2.0 的 GPU 版本（把下面代码保存到 test.py，使用 TF_2G Python 运行），代码如下。

```
import tensorflow as tf
tfabab.test.is_gpu_available()
```

（8）如果输出"True"，则表示 GPU 版本安装成功。

本章介绍了如何在 Linux 和 Windows 系统中安装 TensorFlow 2.0 的 CPU 版本和 GPU 版本。在后面的章节中将使用本章安装的版本对 TensorFlow 2.0 进行介绍。

第 3 章 TensorFlow 的基础概念

本章对 TensorFlow 的基础概念进行简单介绍。

3.1 张量

TensorFlow 包含构建数据流图与计算数据流图等，其构图的基础单元是 Tensors（张量）。张量是 TensorFlow 最核心的组件，所有的运算和优化都是基于张量完成的。

本节将使用 TensorFlow 2.0 对张量进行介绍。

（1）在 TensorFlow 2.0 中，所有的操作都必须导入 TensorFlow，代码如下。

```
#导入 TensorFlow
from __future__ import absolute_import, division, print_function
import tensorflow as tf
#打印 TensorFlow 版本
print(tf.__version__)
```

（2）代码的运行结果是打印目前使用的 TensorFlow 版本。

```
2.0.0-rc2                                            #显示 TensorFlow 版本
```

说明：从 TensorFlow 2.0 开始，默认启用 Eager 模式。TensorFlow 的 Eager 模式是一个命令式，是由运行定义的接口，一旦被调用，其操作立即被执行，无须事先构建静态图。

（3）张量是一个多维数组，在 TensorFlow 2.0 中表现为 tf.Tensor 对象与 NumPy ndarray 对象类似，tf.Tensor 对象具有数据类型和形状。在 TensorFlow 2.0 中，tf.Tensors 可以保持在 GPU 中。TensorFlow 提供了丰富的操作库（tf.add、tf.matmul、tf.linalg.inv 等），灵活使用这些操作库可以方便地操作 tf.Tensor 对象，节省建模时间。

```
#导入 TensorFlow
from __future__ import absolute_import, division, print_function
import tensorflow as tf
#从 NumPy 中导入 int32
from numpy import int32
#打印对应的操作库
print(tf.add(1,2))
print(tf.add([3,8], [2,5]))
print(tf.square(6))
print(tf.reduce_sum([7,8,9]))
print(tf.square(3)+tf.square(4))
```

（4）代码的运行结果如下。

```
tf.Tensor(3, shape=(), dtype=int32)
tf.Tensor([5 13], shape=(2,), dtype=int32)
tf.Tensor(36, shape=(), dtype=int32)
```

```
tf.Tensor(24, shape=(), dtype=int32)
tf.Tensor(25, shape=(), dtype=int32)
```

(5) 在 TensorFlow 2.0 中的每个 Tensor 都有形状和类型，可以通过如下代码进行验证。

```
#导入 TensorFlow
from __future__ import absolute_import, division, print_function
import tensorflow as tf
#从 NumPy 中导入 int32
from numpy import int32
#设置一个有形状的变量
x = tf.matmul([[3], [6]], [[2]])
print(x)
print(x.shape)
print(x.dtype)
```

(6) 代码的运行结果如下。

```
tf.Tensor(
[[6]
 [12]], shape=(2, 1), dtype=int32)
(2, 1)
<dtype: 'int32'>
```

(7) 在 TensorFlow 2.0 中，NumPy 数组和 tf.Tensors 之间最明显的区别是：①张量可以由 GPU（或 TPU）支持；②张量不可变；③TensorFlow tf.Tensors 和 NumPy ndarray 之间的转换很容易。使用下面的例子进行介绍。

```
#导入 TensorFlow 及 NumPy
import tensorflow as tf
import numpy as np
#定义一个 NumPy 数组
ndarray = np.ones([2,2])
#将数组转换成张量
tensor = tf.multiply(ndarray, 36)
print(tensor)
#用 np.add 对 TensorFlow 进行加运算
print(np.add(tensor, 1))
#转换为 NumPy 类型
print(tensor.numpy())
```

(8) 代码的运行结果如下。

```
tf.Tensor(
[[36. 36.]
 [36. 36.]], shape=(2, 2), dtype=float64)
[[37. 37.]
 [37. 37.]]
[[36. 36.]
 [36. 36.]]
```

说明：TensorFlow 操作能够自动将 NumPy ndarray 转换为 Tensors。

3.2 GPU 加速

使用 GPU 可以加速许多 TensorFlow 操作。如果没有任何注释，TensorFlow 会自动决定是使用 GPU 还是 CPU 进行操作。

（1）张量可以在 CPU 和 GPU 内存之间进行复制，其代码如下。

```
#导入 TensorFlow
import tensorflow as tf
#定义一个随机变量
x = tf.random.uniform([3, 3])

print('Is GPU available:')
print(tf.test.is_gpu_available())
print('Is the Tensor on gpu #0:')
print(x.device.endswith('GPU:0'))
```

（2）代码的运行结果如下。

```
Is GPU available:
False
Is the Tensor on gpu #0:
False
```

说明：Tensor.device 提供托管张量内容的设备的完全限定字符串名称。该名称编码了许多详细信息，是分布式执行 TensorFlow 程序所必需的。

（3）在 TensorFlow 中，Placement 指如何分配设备以执行各操作。如果没有明确指定，TensorFlow 会自动决定执行操作的设备，并在需要时将张量复制到该设备。也可以使用 tf.device 上下文管理器将 TensorFlow 操作显式分配到特定设备上，代码如下。

```
#导入 TensorFlow 及 Time
import tensorflow as tf
import time
#定义一个基于时间的函数
def time_matmul(x):
    start = time.time()
#获取当前时间
#在 10 以内循环
    for loop in range(10):
        tf.matmul(x, x)
    result = time.time() - start
    print('10 loops: {:0.2}ms'.format(1000*result))
#强制使用 CPU
print('On CPU:')
with tf.device('CPU:0'):
    x = tf.random.uniform([1000, 1000])
```

```
        #使用断言验证当前是否为CPU0
        assert x.device.endswith('CPU:0')
        time_matmul(x)
    #如果存在GPU,强制使用GPU
    if tf.test.is_gpu_available():
        print('On GPU:')
        with tf.device.endswith('GPU:0'):
            x = tf.random.uniform([1000, 1000])
        #使用断言验证当前是否为GPU0
        assert x.device.endswith('GPU:0')
        time_matmul(x)
```

（4）代码的运行结果如下。

```
10 loops: 2.3e+02ms
```

3.3 数据集

数据集是数据的集合，是机器学习的基础，本节使用 tf.data.Dataset API 构建管道，为模型提供数据。

1. 创建源数据集

在使用数据集之前，需要先创建一个源数据集，使用工厂函数（如 Dataset.from_tensors、Dataset.from_tensor_slices）或从 TextLineDataset 和 TFRecordDataset 等文件中读取的对象来创建源数据集，代码如下。

```
#导入TensorFlow
import tensorflow as tf
#从列表中获取Tensor
ds_tensors = tf.data.Dataset.from_tensor_slices([6,5,4,3,2,1])
#创建csv文件
import tempfile
_, filename = tempfile.mkstemp()
print(filename)
#循环打开文件，获得内容
with open(filename, 'w') as f:
    f.write(" " "Line 1
Line 2
Line 3" " ")
#获取TextLineDataset数据集实例
ds_file = tf.data.TextLineDataset(filename)
```

代码的运行结果如下。

```
/var/folders/tw/0htspm755jj1_vbd_x_zm99w0000gn/T/tmp9tb7kwer
```

打开返回的文件，如下。

```
Line 1
Line 2
Line 3
```

2. 转换函数

将 map、batch 和 shuffle 等转换函数应用于数据集记录。这里以 map 和 shuffle 函数为例，代码如下。

```
#导入 TensorFlow
import tensorflow as tf
#从列表中获取 Tensor
ds_tensors = tf.data.Dataset.from_tensor_slices([6,5,4,3,2,1])
#创建 csv 文件
import tempfile
_, filename = tempfile.mkstemp()
print(filename)
#循环打开文件
with open(filename, 'w') as f:
    f.write(" " "Line 1
Line 2
Line 3" " ")
#获取 TextLineDataset 数据集实例
ds_file = tf.data.TextLineDataset(filename)
ds_tensors = ds_tensors.map(tf.square).shuffle(2).batch(2)
ds_file = ds_file.batch(2)
print (ds_tensors)
print (ds_file)
```

代码的运行结果如下。

```
<BatchDataset shapes: (None,), types: tf.int32>
<BatchDataset shapes: (None,), types: tf.string>
```

3. 迭代

迭代是处理数据集的常用方法，tf.data.Dataset 对象支持迭代循环记录，示例代码如下。

```
#导入 TensorFlow
import tensorflow as tf
#从列表中获取 Tensor
ds_tensors = tf.data.Dataset.from_tensor_slices([6,5,4,3,2,1])
#创建 csv 文件
import tempfile
_, filename = tempfile.mkstemp()
print(filename)
#循环打开文件，获得内容
with open(filename, 'w') as f:
    f.write(" " "Line 1
Line 2
```

```
      Line 3 " " ")
#获取 TextLineDataset 数据集实例
ds_file = tf.data.TextLineDataset(filename)
ds_tensors = ds_tensors.map(tf.square).shuffle(2).batch(2)
ds_file = ds_file.batch(2)
print('ds_tensors 中的元素：')
for x in ds_tensors:
    print(x)
print('ds_file 中的元素：')
for x in ds_file:
    print(x)
```

代码的运行结果如下。

```
ds_tensors 中的元素：
tf.Tensor([36 25], shape=(2,), dtype=int32)
tf.Tensor([9 16], shape=(2,), dtype=int32)
tf.Tensor([1 4], shape=(2,), dtype=int32)
ds_file 中的元素：
tf.Tensor([b'Line 1' b'Line 2'], shape=(2,), dtype=string)
tf.Tensor([b'Line 3'], shape=(1,), dtype=string)
```

本节使用几个简单的例子对数据集进行了初步的介绍。在后面的章节中，会经常使用数据集。

3.4 自定义层

机器学习模型通常可以表示为简单网络层的堆叠与组合，TensorFlow 提供了常见的网络层，TensorFlow 2.0 推荐使用 tf.keras 来构建网络层，以提高可读性和易用性。

3.4.1 网络层的常见操作

TensorFlow 2.0 推荐将 tf.keras 作为构建神经网络的高级 API。本节对常见的网络层操作进行介绍。

（1）构建一个简单的全连接网络，代码如下。

```
#导入 TensorFlow
import tensorflow as tf
#指定网络的神经元个数
layer = tf.keras.layers.Dense(100)
#添加输入维度限制
layer = tf.keras.layers.Dense(100, input_shape=(None, 20))
#每层都可以作为一个函数，将输入的数据作为函数的输入
layer(tf.ones([6, 6]))
print(layer.variables)
```

（2）代码的运行结果如下，包含权重和偏置信息。

```
[<tf.Variable 'dense_1/kernel:0' shape=(6, 100) dtype=float32, numpy=
array([[-0.06826358,  0.13057731,  0.20192285,  0.1940025 ,  0.06154729,
...
        -0.18839078, -0.1510858 , -0.11723022, -0.02339667, -0.166789  ,
        -0.20208013, -0.00988553,  0.1350063 , -0.10279569,  0.12369879]],
      dtype=float32)>, <tf.Variable 'dense_1/bias:0' shape=(100,) dtype=float32, numpy=
array([0., 0., 0., 0., 0., 0., 0., 0., 0., 0., 0., 0., 0., 0., 0., 0.,
       0., 0., 0., 0., 0., 0., 0., 0., 0., 0., 0., 0., 0., 0., 0., 0.,
       0., 0., 0., 0., 0., 0., 0., 0., 0., 0., 0., 0., 0., 0., 0., 0.,
       0., 0., 0., 0., 0., 0., 0., 0., 0., 0., 0., 0., 0., 0., 0., 0.,
       0., 0., 0., 0., 0., 0., 0., 0., 0., 0., 0., 0., 0., 0., 0., 0.,
       0., 0., 0., 0., 0., 0., 0., 0., 0., 0., 0., 0., 0., 0., 0., 0.,
       0., 0., 0., 0.],
      dtype=float32)>]
```

（3）tf.keras 非常灵活，还可以分别取出上例中的权重和偏置，代码如下。

```
#导入 TensorFlow
import tensorflow as tf
#指定网络的神经元个数
layer = tf.keras.layers.Dense(100)
#添加输入维度限制
layer = tf.keras.layers.Dense(100, input_shape=(None, 20))
#每层都可以作为一个函数，将输入数据作为函数的输入
layer(tf.ones([6, 6]))
#分别取出权重和偏置
print(layer.kernel, layer.bias)
```

（4）代码的运行结果如下。

```
<tf.Variable 'dense_1/kernel:0' shape=(6, 100) dtype=float32, numpy=
array([[-7.21279234e-02, -1.41769499e-02, -9.59615260e-02,
        -9.15659666e-02, -1.70545742e-01, -6.98231310e-02,
        -1.63669288e-02, -2.02601537e-01,  6.32829815e-02,
...
        -1.99535131e-01, -1.68282419e-01,  1.06709614e-01,
        -1.87540770e-01]], dtype=float32)> <tf.Variable 'dense_1/bias:0' shape=(100,) dtype=float32, numpy=
array([0., 0., 0., 0., 0., 0., 0., 0., 0., 0., 0., 0., 0., 0., 0., 0.,
       0., 0., 0., 0., 0., 0., 0., 0., 0., 0., 0., 0., 0., 0., 0., 0.,
       0., 0., 0., 0., 0., 0., 0., 0., 0., 0., 0., 0., 0., 0., 0., 0.,
       0., 0., 0., 0., 0., 0., 0., 0., 0., 0., 0., 0., 0., 0., 0., 0.,
       0., 0., 0., 0., 0., 0., 0., 0., 0., 0., 0., 0., 0., 0., 0., 0.,
```

3.4.2 自定义网络层

在实际中，经常需要扩展 tf.keras.Layer 类并自定义网络层。本节介绍如何自定义网络层。
（1）在自定义网络层的过程中主要使用 3 个函数，示例代码如下。

```
#导入 TensorFlow
import tensorflow as tf
#定义一个基础网络类
class MyDense(tf.keras.layers.Layer):
    #初始化函数
    def __init__(self, n_outputs):
        super(MyDense, self).__init__()
        self.n_outputs = n_outputs
    #build()函数，获得输入张量的形状，并可以进行其余的初始化
    def build(self, input_shape):
        self.kernel = self.add_weight('kernel', shape=[int(input_shape[-1]),
            self.n_outputs])
    #call()函数，构建网络结构，进行前向传播
    def call(self, input):
        return tf.matmul(input, self.kernel)
#使用基础网络类构建网络层
layer = MyDense(10)
#注释说明打印输出的是什么
print(layer(tf.ones([6, 5])))
print(layer.trainable_variables)
```

说明：调用 build()函数构建网络并不是必要的，有时可以在__init__()中构建网络。但是，调用 build()函数构建网络的优点是可以动态获取输入数据的 shape，大大提高了运行效率。

（2）上述代码的运行结果如下。

```
tf.Tensor(
[[-0.2683453  -0.28268558  0.8902599   0.49899605 -0.12212557  0.4986639
   0.4618655  -0.6703784   1.413507   -0.03647304]
 ...
 [-0.2683453  -0.28268558  0.8902599   0.49899605 -0.12212557  0.4986639
   0.4618655  -0.6703784   1.413507   -0.03647304]], shape=(6, 10), dtype=float32)
[<tf.Variable 'my_dense/kernel:0' shape=(5, 10) dtype=float32, numpy=
array([[ 0.10833806, -0.36234528,  0.2948733 , -0.07032776, -0.42154115,
         0.4700827 ,  0.4416079 , -0.34874937,  0.59902436, -0.15988943],
       [-0.5554653 ,  0.06153011,  0.4769748 ,  0.43647486,  0.5021389 ,
        -0.2384331 ,  0.12493771, -0.11590213,  0.1238513 , -0.39060408],
       [-0.19530559, -0.5491605 , -0.40542966, -0.37616614, -0.32701793,
         0.59960526, -0.1215685 , -0.24531192,  0.45543557, -0.5302647 ],
```

```
       [-0.09493732,  0.25326592,  0.19607717,  0.11132509,  0.46998113,
        -0.40997064, -0.49282327, -0.38274086,  0.36849576,  0.5851895 ],
       [ 0.46902484,  0.31402415,  0.32776433,  0.39769   , -0.34568653,
         0.0773797 ,  0.5097117 ,  0.4223259 , -0.1333    ,  0.45909554]],
      dtype=float32)>]
```

3.4.3 网络层组合

有很多机器学习模型是不同网络层的组合,网络层组合学习是加快学习速度和精度的重要方法。

(1) 使用下面的代码在 TensorFlow 2.0 中构建一个包含多个网络层的模型。

```
#导入 TensorFlow
import tensorflow as tf
#残差块
class ResnetBlock(tf.keras.Model):
    def __init__(self, kernel_size, filters):
        super(ResnetBlock, self).__init__(name='resnet_block')
        filter1, filter2, filter3 = filters
        #共 3 个子层,每个子层有 1 个卷积和 1 个批正则化
        #第 1 个子层,1*1 的卷积
        self.conv1 = tf.keras.layers.Conv2D(filter1, (1,1))
        self.bn1 = tf.keras.layers.BatchNormalization()
        #第 2 个子层,使用 kernel_size
        self.conv2 = tf.keras.layers.Conv2D(filter2, kernel_size, padding='same')
        self.bn2 = tf.keras.layers.BatchNormalization()
        #第 3 个子层,1*1 的卷积
        self.conv3 = tf.keras.layers.Conv2D(filter3, (1,1))
        self.bn3 = tf.keras.layers.BatchNormalization()
    def call(self, inputs, training=False):
        #堆叠每个子层
        x = self.conv1(inputs)
        x = self.bn1(x, training=training)
        x = self.conv2(x)
        x = self.bn2(x, training=training)
        x = self.conv3(x)
        x = self.bn3(x, training=training)
        #残差连接
        x += inputs
        outputs = tf.nn.relu(x)
        return outputs
resnetBlock = ResnetBlock(2, [6,4,9])
#数据测试
```

```
print(resnetBlock(tf.ones([1,3,9,9])))
#查看网络中的变量名
print([x.name for x in resnetBlock.trainable_variables])
```

说明：该例子是 resnet 的一个残差块，是"卷积+批标准化+残差连接"的组合。

（2）代码的运行结果如下。

```
tf.Tensor(
[[[[0.         0.28857142 0.07571369 1.1377857  2.2751708  1.4268001
    1.3483192  1.4295883  0.        ]
  ...
   [0.5324651  0.84801346 1.1660165  0.91560936 1.2024628  0.9481153
    1.0713446  1.0192398  1.3742731 ]]]], shape=(1, 3, 9, 9), dtype=float32)
```

（3）有时需要构建线性模型，可以直接用 tf.keras.Sequential 来构建，示例代码如下。

```
#导入 TensorFlow
import tensorflow as tf
#构建一个 3 层的线性模型，使用默认参数
seq_model = tf.keras.Sequential(
[
    tf.keras.layers.Conv2D(1, 1, input_shape=(None, None, 3)),
    tf.keras.layers.BatchNormalization(),
    tf.keras.layers.Conv2D(2, 1, padding='same'),
    tf.keras.layers.BatchNormalization(),
    tf.keras.layers.Conv2D(3, 1),
    tf.keras.layers.BatchNormalization(),
])
seq_modelBlock = seq_model(tf.ones([1,2,3,3]))
print (seq_modelBlock)
```

（4）代码的运行结果如下。

```
tf.Tensor(
[[[[-0.0002618  -0.04231375 -0.02505807]
   [-0.0002618  -0.04231375 -0.02505807]
   [-0.0002618  -0.04231375 -0.02505807]]
  ...
  [[-0.0002618  -0.04231375 -0.02505807]
   [-0.0002618  -0.04231375 -0.02505807]
   [-0.0002618  -0.04231375 -0.02505807]]]], shape=(1, 2, 3, 3), dtype=float32)
```

3.4.4 自动求导

TensorFlow 使用的求导方法被称为自动微分，它既不是符号求导也不是数值求导，而是两者的结合。

（1）TensorFlow 2.0 利用 tf.GradientTape API 来实现自动求导功能，在 tf.GradientTape()上

下文中执行的操作都会被记录在"tape"中，然后 TensorFlow 2.0 使用反向自动微分来计算相关操作的梯度，示例代码如下。

```
#导入 TensorFlow
import tensorflow as tf
#构建数据集
x = tf.ones((2,2))
#需要计算梯度的操作
with tf.GradientTape() as t:
    t.watch(x)
    y = tf.reduce_sum(x)
    z = tf.multiply(y,y)
#计算 z 关于 x 的梯度
dz_dx = t.gradient(z, x)
print(dz_dx)
```

（2）代码的运行结果如下。

```
tf.Tensor(
[[8. 8.]
 [8. 8.]], shape=(2, 2), dtype=float32)
```

（3）输出中间变量的导数，代码如下。

```
#导入 TensorFlow
import tensorflow as tf
#构建数据集
x = tf.ones((2,2))
#每个 tape 只能进行一次梯度求导
with tf.GradientTape() as t:
    t.watch(x)
    y = tf.reduce_sum(x)
    z = tf.multiply(y,y)
dz_dy = t.gradient(z, y)
print(dz_dy)
```

（4）代码的运行结果如下。

```
tf.Tensor(8.0, shape=(), dtype=float32)
```

（5）在默认情况下，GradientTape 的资源会在执行 tf.GradientTape()后释放。如果希望多次计算梯度，需要创建一个持久的 GradientTape。代码如下。

```
#导入 TensorFlow
import tensorflow as tf
#构建数据集
x = tf.ones((2,2))
#创建持久的 GradientTape
with tf.GradientTape(persistent=True) as t:
    t.watch(x)
    y = tf.reduce_sum(x)
```

```
        z = tf.multiply(y, y)
    #使用创建的 GradientTape 进行计算
    dz_dx = t.gradient(z, x)
    print(dz_dx)
    dz_dy = t.gradient(z, y)
    print(dz_dy)
```

(6) 代码的运行结果如下。

```
tf.Tensor(
[[8. 8.]
 [8. 8.]], shape=(2, 2), dtype=float32)
tf.Tensor(8.0, shape=(), dtype=float32)
```

(7) 因为 tape 记录了整个操作,所以即使存在 Python 控制流(如 if 和 while),也能正常处理梯度求导,示例代码如下。

```
#导入 TensorFlow
import tensorflow as tf
def f(x, y):
    output = 1.0
    #循环
    for i in range(y):
        #对每个项进行判断
        if i>1 and i<5:
            output = tf.multiply(output, x)
    return output
def grad(x, y):
    with tf.GradientTape() as t:
        t.watch(x)
        out = f(x, y)
        #返回梯度
        return t.gradient(out, x)
#x 为固定值
x = tf.convert_to_tensor(2.0)
print(grad(x, 6))
print(grad(x, 5))
print(grad(x, 4))
```

(8) 代码的运行结果如下。

```
tf.Tensor(12.0, shape=(), dtype=float32)
tf.Tensor(12.0, shape=(), dtype=float32)
tf.Tensor(4.0, shape=(), dtype=float32)
```

(9) GradientTape 上下文管理器在计算梯度时会保持梯度。因此,GradientTape 也可以实现高阶梯度计算,示例代码如下。

```
#导入 TensorFlow
import tensorflow as tf
#使用 Variable 构建实验数据
```

```
x = tf.Variable(1.0)
with tf.GradientTape() as t1:
    with tf.GradientTape() as t2:
        y = x * x * x
    dy_dx = t2.gradient(y, x)
    print(dy_dx)
d2y_d2x = t1.gradient(dy_dx, x)
print(d2y_d2x)
```

（10）代码的运行结果如下。

```
tf.Tensor(3.0, shape=(), dtype=float32)
tf.Tensor(6.0, shape=(), dtype=float32)
```

本章介绍了一些 TensorFlow 2.0 的基本概念并举例进行了说明，后面在使用到这些概念时，会举例对具体的使用方法进行介绍。

第4章 TensorFlow 与多层感知器

多层感知器（MLP）是一种前馈人工神经网络模型，可以将输入的多个数据集映射到单一的输出数据集上。

在前面的章节中已经介绍了如何使用 TensorFlow 2.0 构建单层的神经网络。但是仅使用单层的神经网络并不能保证模型的准确率，这时需要使用多层感知器构建模型以保证模型的准确率。

4.1 MLP 简介

MLP 是一种趋向结构的人工神经网络，映射一组输入向量到一组输出向量。可以将 MLP 看作由多个节点层组成的有向图，每一层全连接到下一层。通常使用反向传播算法的监督学习方法训练 MLP。

4.2 基础 MLP 网络

本节使用回归分析和分类任务两个机器学习的典型应用场景对基础 MLP 网络进行介绍。

4.2.1 回归分析

回归分析是确定两种或两种以上变量相互依赖的定量关系的统计分析方法，本节使用 TensorFlow 2.0 对回归分析进行介绍。

（1）导入数据集，代码如下。

```
#以 boston_housing 数据集为基础构建模型
from tensorflow import keras
x_train, y_train), (x_test, y_test) = keras.datasets.boston_housing.load_data()
#打印相应结果进行校验
print(x_train.shape, ' ', y_train.shape)
print(x_test.shape, ' ', y_test.shape)
```

（2）代码的运行结果如下。

```
(404, 13), ' ', (404,)
((102, 13), ' ', (102,))
```

（3）在导入数据正确的前提下，构建并配置回归分析模型，代码如下。

```
#以 TensorFlow 为基础构建 Keras
from tensorflow import keras
from tensorflow.keras import layers
```

```
(x_train, y_train), (x_test, y_test) = keras.datasets.boston_housing.load_data()
#以 boston_housing 数据集为基础构建模型
model = keras.Sequential([
    layers.Dense(32, activation='sigmoid', input_shape=(13,)),
    layers.Dense(32, activation='sigmoid'),
    layers.Dense(32, activation='sigmoid'),
    layers.Dense(1)
])
#使用密集连接层类构建模型
model.compile(optimizer=keras.optimizers.SGD(0.1),
    loss='mean_squared_error',  #keras.losses.mean_squared_error
    metrics=['mse'])
#编译模型
model.summary()
#输出各层的情况
```

(4) 代码的运行结果如下。

```
Model: " sequential "
_____
Layer (type)                 Output Shape              Param #
=================================================================
dense (Dense)                (None, 32)                448
_____
dense_1 (Dense)              (None, 32)                1056
_____
dense_2 (Dense)              (None, 32)                1056
_____
dense_3 (Dense)              (None, 1)                 33
=================================================================
Total params: 2,593
Trainable params: 2,593
Non-trainable params: 0
_____
```

(5) 结果显示模型输出正确。下面对回归分析模型进行训练,本例中的训练次数为 50 次,代码如下。

```
#以 TensorFlow 为基础构建 Keras
from tensorflow import keras
from tensorflow.keras import layers
(x_train, y_train), (x_test, y_test) = keras.datasets.boston_housing.load_data()
#以 boston_housing 数据集为基础构建模型
model = keras.Sequential([
    layers.Dense(32, activation='sigmoid', input_shape=(13,)),
    layers.Dense(32, activation='sigmoid'),
    layers.Dense(32, activation='sigmoid'),
```

```
    layers.Dense(1)
])
#使用密集连接层类构建模型
model.compile(optimizer=keras.optimizers.SGD(0.1),
    loss='mean_squared_error',  #keras.losses.mean_squared_error
    metrics=['mse'])
#编译模型
model.fit(x_train, y_train, batch_size=50, epochs=50, validation_split=0.1, verbose=1)
#迭代训练模型
```

（6）代码的运行结果如下。

```
Train on 363 samples, validate on 41 samples
Epoch 1/50
...
363/363 [==============================] - 0s 24us/sample - loss: 86.7216 - mse: 86.7216 - val_loss: 40.2732 - val_mse: 40.2732
Epoch 50/50
363/363 [==============================] - 0s 24us/sample - loss: 83.4134 - mse: 83.4134 - val_loss: 33.5665 - val_mse: 33.5665
```

（7）对模型进行多次训练后，集中输出训练结果，代码如下。

```
#以 TensorFlow 为基础构建 Keras
from tensorflow import keras
from tensorflow.keras import layers
#以 boston_housing 数据集为基础构建模型
(x_train, y_train), (x_test, y_test) = keras.datasets.boston_housing.load_data()
#使用 sigmoid 算法构建一个线性时序模型
model = keras.Sequential([
    layers.Dense(32, activation='sigmoid', input_shape=(13,)),
    layers.Dense(32, activation='sigmoid'),
    layers.Dense(32, activation='sigmoid'),
    layers.Dense(1)
])
#使用密集连接层类构建模型
model.compile(optimizer=keras.optimizers.SGD(0.1),
    loss='mean_squared_error',  # keras.losses.mean_squared_error
    metrics=['mse'])
#编译模型
model.fit(x_train, y_train, batch_size=50, epochs=50, validation_split=0.1, verbose=1)
#迭代训练模型
result = model.evaluate(x_test, y_test)
#定义返回结果
print(model.metrics_names)
print(result)
```

(8）代码的运行结果如下。

```
Epoch 1/50
363/363 [==============================] - 0s 593us/sample - loss: 466.8626 - mse: 466.8626 - val_loss: 83.5865 - val_mse: 83.5865
...
Epoch 50/50
363/363 [==============================] - 0s 22us/sample - loss: 90.1021 - mse: 90.1021 - val_loss: 42.5076 - val_mse: 42.5076
102/102 [==============================] - 0s 20us/sample - loss: 87.3885 - mse: 87.3885
['loss', 'mse']
[87.38852317660462, 87.38853]
```

4.2.2 分类任务

分类任务通过训练一个特定的函数来判断输入数据所属的类别。分类任务在现实中的应用非常广泛，如图像鉴定、语音识别等。

（1）导入数据集，代码如下。

```
#使用 TensorFlow 内置数据集
from sklearn.datasets import load_breast_cancer
from sklearn.model_selection import train_test_split

whole_data = load_breast_cancer()
x_data = whole_data.data
y_data = whole_data.target
#以 breast_cancer 数据集为基础构建模型
x_train, x_test, y_train, y_test = train_test_split(x_data, y_data, test_size=0.3, random_state=7)
print(x_train.shape, ' ', y_train.shape)
print(x_test.shape, ' ', y_test.shape)
```

（2）得到如下结果则说明导入数据正确。

```
((398, 30), ' ', (398,))
((171, 30), ' ', (171,))
```

（3）在导入数据正确的前提下，构建并配置分类任务模型，代码如下。

```
#以 TensorFlow 为基础构建 Keras 并导入数据集
from tensorflow import keras
from tensorflow.keras import layers
from sklearn.datasets import load_breast_cancer
from sklearn.model_selection import train_test_split
whole_data = load_breast_cancer()
x_data = whole_data.data
y_data = whole_data.target
#以 breast_cancer 数据集为基础构建模型
```

```
        x_train, x_test, y_train, y_test = train_test_split(x_data, y_data, test_size=0.3,
random_state=7)
        model = keras.Sequential([
            layers.Dense(32, activation='relu', input_shape=(30,)),
            layers.Dense(32, activation='relu'),
            layers.Dense(1, activation='sigmoid')
        ])
        #使用密集连接层类构建模型
        model.compile(optimizer=keras.optimizers.Adam(),
            loss=keras.losses.binary_crossentropy,
            metrics=['accuracy'])
        #编译模型
        model.summary()
        #输出各层的情况
```

（4）对模型进行校验，结果如下。

```
Model: "sequential"
_____
Layer (type)                 Output Shape              Param #
=================================================================
dense (Dense)                (None, 32)                992
_____
dense_1 (Dense)              (None, 32)                1056
_____
dense_2 (Dense)              (None, 1)                 33
=================================================================
Total params: 2,081
Trainable params: 2,081
Non-trainable params: 0
```

（5）在模型输出正确的前提下，对回归分析模型进行训练，本例中的训练次数为 10 次，代码如下。

```
        #以 TensorFlow 为基础构建 Keras 并导入数据集
        from tensorflow import keras
        from tensorflow.keras import layers
        from sklearn.datasets import load_breast_cancer
        from sklearn.model_selection import train_test_split
        whole_data = load_breast_cancer()
        x_data = whole_data.data
        y_data = whole_data.target
        #以 breast_cancer 数据集为基础构建模型
        x_train, x_test, y_train, y_test = train_test_split(x_data, y_data, test_size=0.3,
random_state=7)
        model = keras.Sequential([
            layers.Dense(32, activation='relu', input_shape=(30,)),
```

```
        layers.Dense(32, activation='relu'),
        layers.Dense(1, activation='sigmoid')
    ])
    #使用密集连接层类构建模型
    model.compile(optimizer=keras.optimizers.Adam(),
        loss=keras.losses.binary_crossentropy,
        metrics=['accuracy'])
    #编译模型
    model.fit(x_train, y_train, batch_size=64, epochs=10, verbose=1)
    #制订训练计划,训练10次
    model.evaluate(x_test, y_test)
```

(6) 代码的运行结果如下。

```
Epoch 1/10
398/398 [==============================] - 0s 460us/sample - loss: 121.5031 - accuracy: 0.6055
...
Epoch 10/10
398/398 [==============================] - 0s 18us/sample - loss: 0.4765 - accuracy: 0.8090
171/171 [==============================] - 0s 263us/sample - loss: 0.4277 - accuracy: 0.8538
```

(7) 对模型进行多次训练后,集中输出训练结果,代码如下。

```
#以 TensorFlow 为基础构建 Keras 并导入数据集
from tensorflow import keras
from tensorflow.keras import layers
from sklearn.datasets import load_breast_cancer
from sklearn.model_selection import train_test_split
whole_data = load_breast_cancer()
x_data = whole_data.data
y_data = whole_data.target
#以 breast_cancer 数据集为基础构建模型
x_train, x_test, y_train, y_test = train_test_split(x_data, y_data, test_size=0.3, random_state=7)
model = keras.Sequential([
    layers.Dense(32, activation='relu', input_shape=(30,)),
    layers.Dense(32, activation='relu'),
    layers.Dense(1, activation='sigmoid')
])
#使用密集连接层类构建模型
model.compile(optimizer=keras.optimizers.Adam(),
    loss=keras.losses.binary_crossentropy,
    metrics=['accuracy'])
#编译模型
```

```
model.fit(x_train, y_train, batch_size=64, epochs=10, verbose=1)
#制订训练计划,训练10次
model.evaluate(x_test, y_test)
#输入数据和标签,输出损失率和精确度
print(model.metrics_names)
```

(8)代码的运行结果如下。

```
Epoch 1/10
398/398 [==============================] - 0s 451us/sample - loss: 59.2639 - accuracy: 0.6030
...
Epoch 10/10
398/398 [==============================] - 0s 16us/sample - loss: 2.2104 - accuracy: 0.3342
171/171 [==============================] - 0s 281us/sample - loss: 1.3726 - accuracy: 0.4503
['loss', 'accuracy']
```

本例使用 breast_cancer 的数据源进行了简单的图像分类演示,可以发现,随着训练次数的增加,损失率(loss)不断下降,而精确度(accuracy)不断上升。这就是机器学习进行多次训练的意义。

4.3 基础模型

本节构建一个基础模型,以与后面章节中的优化模型进行对比。

(1)导入数据集,代码如下。

```
#以 TensorFlow 为基础构建 Keras 并导入数据集
from tensorflow import keras
from tensorflow.keras import layers
(x_train, y_train), (x_test, y_test) = keras.datasets.mnist.load_data()
x_train = x_train.reshape([x_train.shape[0], -1])
x_test = x_test.reshape([x_test.shape[0], -1])
#导入 MNIST 数据集
print(x_train.shape, ' ', y_train.shape)
print(x_test.shape, ' ', y_test.shape)
```

(2)代码的运行结果如下。

```
(60000, 784), ' ', (60000,)
((10000, 784), ' ', (10000,))
```

(3)在导入数据正确的前提下,构建并配置模型,代码如下。

```
#以 TensorFlow 为基础构建 Keras 并导入数据集
from tensorflow import keras
from tensorflow.keras import layers
```

```
(x_train, y_train), (x_test, y_test) = keras.datasets.mnist.load_data()
x_train = x_train.reshape([x_train.shape[0], -1])
x_test = x_test.reshape([x_test.shape[0], -1])
#导入MNIST数据集
model = keras.Sequential([
    layers.Dense(64, activation='relu', input_shape=(784,)),
    layers.Dense(64, activation='relu'),
    layers.Dense(64, activation='relu'),
    layers.Dense(10, activation='softmax')
])
#构建一个不添加任何条件的模型
model.compile(optimizer=keras.optimizers.Adam(),
    loss=keras.losses.SparseCategoricalCrossentropy(),
    metrics=['accuracy'])
#编译模型
model.summary()
#打印各层情况
```

（4）代码的运行结果如下。

```
Model: "sequential"
_____
Layer (type)                 Output Shape              Param #
=================================================================
dense (Dense)                (None, 64)                50240
_____
dense_1 (Dense)              (None, 64)                4160
_____
dense_2 (Dense)              (None, 64)                4160
_____
dense_3 (Dense)              (None, 10)                650
=================================================================
Total params: 59,210
Trainable params: 59,210
Non-trainable params: 0
_____
```

（5）在模型输出正确的前提下，对回归分析模型进行训练，本例中的训练次数为100次，显示图像，代码如下。

```
#以TensorFlow为基础构建Keras并导入数据集
from tensorflow import keras
from tensorflow.keras import layers
(x_train, y_train), (x_test, y_test) = keras.datasets.mnist.load_data()
x_train = x_train.reshape([x_train.shape[0], -1])
x_test = x_test.reshape([x_test.shape[0], -1])
```

```
#导入MNIST数据集
model = keras.Sequential([
    layers.Dense(64, activation='relu', input_shape=(784,)),
    layers.Dense(64, activation='relu'),
    layers.Dense(64, activation='relu'),
    layers.Dense(10, activation='softmax')
])
#构建一个不添加任何条件的模型
model.compile(optimizer=keras.optimizers.Adam(),
    loss=keras.losses.SparseCategoricalCrossentropy(),
    metrics=['accuracy'])
#编译模型
history = model.fit(x_train, y_train, batch_size=256, epochs=100, validation_split=0.3, verbose=0)
#制订训练计划,训练100次
import matplotlib.pyplot as plt
plt.plot(history.history['accuracy'])
plt.plot(history.history['val_accuracy'])
plt.legend(['training', 'validation'], loc='upper left')
plt.show()
#绘图
```

(6) 得到对比值曲线如图 4-1 所示。

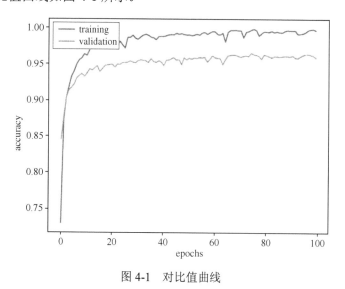

图 4-1 对比值曲线

```
10000/10000 [==============================] - 0s 27us/sample - loss: 0.4660 - accuracy: 0.9611
```

本例构建了一个基础的数据分析模型,在后面的章节中会对相同的数据和模型进行优化并对比输出结果。

4.4 权重初始化

神经网络及深度学习模型训练的本质是对权重进行更新，本节使用 TensorFlow 2.0 对权重初始化进行介绍。

（1）导入数据集，代码如下。

```
from tensorflow import keras
from tensorflow.keras import layers
(x_train, y_train), (x_test, y_test) = keras.datasets.mnist.load_data()
x_train = x_train.reshape([x_train.shape[0], -1])
x_test = x_test.reshape([x_test.shape[0], -1])
#导入 MNIST 数据集
print(x_train.shape, ' ', y_train.shape)
print(x_test.shape, ' ', y_test.shape)
```

（2）代码的运行结果如下。

```
(60000, 784), ' ', (60000,)
((10000, 784), ' ', (10000,))
```

（3）在导入数据正确的前提下，构建并配置相应的模型，代码如下。

```
from tensorflow import keras
from tensorflow.keras import layers
(x_train, y_train), (x_test, y_test) = keras.datasets.mnist.load_data()
x_train = x_train.reshape([x_train.shape[0], -1])
x_test = x_test.reshape([x_test.shape[0], -1])
#导入 MNIST 数据集
model = keras.Sequential([
    layers.Dense(64, activation='relu', kernel_initializer='he_normal', input_shape=(784,)),
    layers.Dense(64, activation='relu', kernel_initializer='he_normal'),
    layers.Dense(64, activation='relu', kernel_initializer='he_normal'),
    layers.Dense(10, activation='softmax')
])
#构建一个添加权重值 kernel_initializer='he_normal'的模型
model.compile(optimizer=keras.optimizers.Adam(),
    loss=keras.losses.SparseCategoricalCrossentropy(),
    metrics=['accuracy'])
#编译模型
model.summary()
#输出各层情况
```

（4）代码的运行结果如下。

```
Model: "sequential"
_____
Layer (type)                 Output Shape              Param #
```

```
=================================================================
dense (Dense)                (None, 64)                50240
_____
dense_1 (Dense)              (None, 64)                4160
_____
dense_2 (Dense)              (None, 64)                4160
_____
dense_3 (Dense)              (None, 10)                650
=================================================================
Total params: 59,210
Trainable params: 59,210
Non-trainable params: 0
_____
```

（5）在模型输出正确的前提下，对回归分析模型进行训练，本例中的训练次数为100次，显示图像，代码如下。

```
from tensorflow import keras
from tensorflow.keras import layers
(x_train, y_train), (x_test, y_test) = keras.datasets.mnist.load_data()
x_train = x_train.reshape([x_train.shape[0], -1])
x_test = x_test.reshape([x_test.shape[0], -1])
#导入MNIST数据集
model = keras.Sequential([
    layers.Dense(64, activation='relu', kernel_initializer='he_normal', input_shape=(784,)),
    layers.Dense(64, activation='relu', kernel_initializer='he_normal'),
    layers.Dense(64, activation='relu', kernel_initializer='he_normal'),
    layers.Dense(10, activation='softmax')
])
#构建一个添加权重值kernel_initializer='he_normal'的模型
model.compile(optimizer=keras.optimizers.Adam(),
     loss=keras.losses.SparseCategoricalCrossentropy(),
     metrics=['accuracy'])
#编译模型
history = model.fit(x_train, y_train, batch_size=256, epochs=100, validation_split= 0.3, verbose=0)
#制订训练计划，训练100次
import matplotlib.pyplot as plt
plt.plot(history.history['accuracy'])
plt.plot(history.history['val_accuracy'])
plt.legend(['training', 'validation'], loc='upper left')
plt.show()
result = model.evaluate(x_test, y_test)
#打印相应的结果
```

（6）代码的运行结果如下，权重初始化的训练结果如图 4-2 所示。

```
10000/10000 [==============================] - 0s 36us/sample - loss: 136.0406 - accuracy: 0.0809
```

本例采用了与基本数据集相同的数据集，权重为 he_normal。通过对比两者的图像和输出结果可以看出，添加权重会对结果产生影响。选择合适的初始权重值对数据分析的结果和学习时间有很大影响。

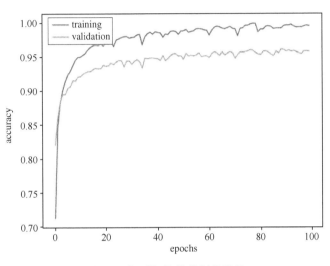

图 4-2　权重初始化的训练结果

说明：he_normal 是 He 正态分布初始化方法，参数由均值为 0、标准差为 sqrt(2 / fan_in) 的正态分布产生，其中 fan_in 为权重张量。

4.5　激活函数

激活函数在运行时，激活神经网络中的某部分神经元，并将激活神经元的信息输入到下一层神经网络中。神经网络之所以能处理非线性问题，是因为激活函数具有非线性表达能力。

（1）本节以 sigmoid 为例进行讲解，构建以 sigmoid 为激活函数的模型，代码如下。

```
#以 TensorFlow 为基础构建 Keras
import ssl
from tensorflow import keras
from tensorflow.keras import layers
#导入图形化工具
import matplotlib.pyplot as plt
#由于是 HTTPS 协议域名，需要添加相应证书
ssl._create_default_https_context = ssl._create_unverified_context
(x_train, y_train), (x_test, y_test) = keras.datasets.mnist.load_data()
x_train = x_train.reshape([x_train.shape[0], -1])
```

```
x_test = x_test.reshape([x_test.shape[0], -1])
#导入 MNIST 数据集
model = keras.Sequential([
    layers.Dense(64, activation='sigmoid', input_shape=(784,)),
    layers.Dense(64, activation='sigmoid'),
    layers.Dense(64, activation='sigmoid'),
    layers.Dense(10, activation='softmax')
])
#构建一个添加激活函数 activation='sigmoid'的模型
model.compile(optimizer=keras.optimizers.Adam(),
    loss=keras.losses.SparseCategoricalCrossentropy(),
    metrics=['accuracy'])
#编译模型
model.summary()
#打印各层情况
```

说明：sigmoid 函数也称为 logistic 函数，用于隐层神经元输出，取值范围为(0,1)，它可以将一个实数映射到(0,1)区间，可以进行二分类。

（2）根据导入的数据和构建的模型，可以得到如下结果。

```
Model: "sequential"
_____
Layer (type)                 Output Shape              Param #
=================================================================
dense (Dense)                (None, 64)                50240
_____
dense_1 (Dense)              (None, 64)                4160
_____
dense_2 (Dense)              (None, 64)                4160
_____
dense_3 (Dense)              (None, 10)                650
=================================================================
Total params: 59,210
Trainable params: 59,210
Non-trainable params: 0
_____
```

（3）训练 100 次后，显示图像，代码如下。

```
#以 TensorFlow 为基础构建 Keras
import ssl
from tensorflow import keras
from tensorflow.keras import layers
#导入图形化工具
import matplotlib.pyplot as plt
#由于是 HTTPS 协议域名，需要添加相应证书
ssl._create_default_https_context = ssl._create_unverified_context
```

```
(x_train, y_train), (x_test, y_test) = keras.datasets.mnist.load_data()
x_train = x_train.reshape([x_train.shape[0], -1])
x_test = x_test.reshape([x_test.shape[0], -1])
#导入MNIST数据集
model = keras.Sequential([
    layers.Dense(64, activation='sigmoid', input_shape=(784,)),
    layers.Dense(64, activation='sigmoid'),
    layers.Dense(64, activation='sigmoid'),
    layers.Dense(10, activation='softmax')
])
model.compile(optimizer=keras.optimizers.Adam(),
    loss=keras.losses.SparseCategoricalCrossentropy(),
    metrics=['accuracy'])
#构建一个添加激活函数activation='sigmoid'的模型
history = model.fit(x_train, y_train, batch_size=256, epochs=100, validation_split= 0.3, verbose=0)
#制订训练计划，训练100次
import matplotlib.pyplot as plt
plt.plot(history.history['accuracy'])
plt.plot(history.history['accuracy'])
plt.plot(history.history['val_accuracy'])
plt.legend(['training', 'validation'], loc='upper left')
plt.show()
result = model.evaluate(x_test, y_test)
#打印输出结果
```

（4）运行代码，得到激活函数的训练结果如图4-3所示。

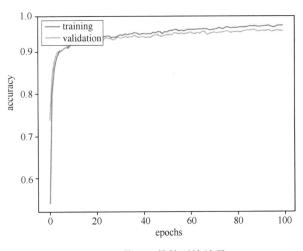

图4-3 激活函数的训练结果

```
10000/10000 [==============================] - 0s 41us/sample - loss: 2.6070 - accuracy: 0.0956
```

本例采用了与基本数据集相同的数据集，定义 sigmoid 为激活函数。通过对比两者的图像和输出结果可以看出，指定激活函数对结果精确度的影响比较明显。说明 sigmoid 激活函数适合用于分析这个数据集。

4.6 批标准化

批标准化要解决的问题是：模型参数在学习阶段的变化会使每个隐藏层输出的分布发生变化。这意味着靠后的层要在训练过程中适应这些变化。批标准化是一种简单、高效的改善神经网络性能的方法。

（1）构建批标准化模型，代码如下。

```
#以 TensorFlow 为基础构建 Keras
import ssl
from tensorflow import keras
from tensorflow.keras import layers
#导入图形化工具
import matplotlib.pyplot as plt
#由于是 HTTPS 协议域名，需要添加相应证书
ssl._create_default_https_context = ssl._create_unverified_context
(x_train, y_train), (x_test, y_test) = keras.datasets.mnist.load_data()
x_train = x_train.reshape([x_train.shape[0], -1])
x_test = x_test.reshape([x_test.shape[0], -1])
#导入 MNIST 数据集
model = keras.Sequential([
    layers.Dense(64, activation='relu', input_shape=(784,)),
    layers.BatchNormalization(),
    layers.Dense(64, activation='relu'),
    layers.BatchNormalization(),
    layers.Dense(64, activation='relu'),
    layers.BatchNormalization(),
    layers.Dense(10, activation='softmax')
])
#对每个层进行标准化，具体参数为 layers.BatchNormalization()
model.compile(optimizer=keras.optimizers.SGD(),
    loss=keras.losses.SparseCategoricalCrossentropy(),
    metrics=['accuracy'])
#编译模型
model.summary()
#打印各层情况
```

（2）根据导入的数据和构建的模型，可以得到如下结果。

```
Model: "sequential"
```

```
Layer (type)                    Output Shape            Param #
=================================================================
dense (Dense)                   (None, 64)              50240
_____
batch_normalization_v2 (Batc    (None, 64)              256
_____
dense_1 (Dense)                 (None, 64)              4160
_____
batch_normalization_v2_1 (Ba    (None, 64)              256
_____
dense_2 (Dense)                 (None, 64)              4160
_____
batch_normalization_v2_2 (Ba    (None, 64)              256
_____
dense_3 (Dense)                 (None, 10)              650
=================================================================
Total params: 59,978
Trainable params: 59,594
Non-trainable params: 384
_____
```

（3）训练100次后，显示图像，代码如下。

```
#以 TensorFlow 为基础构建 Keras
import ssl
from tensorflow import keras
from tensorflow.keras import layers
#导入图形化工具
import matplotlib.pyplot as plt
#由于是 HTTPS 协议域名，需要添加相应证书
ssl._create_default_https_context = ssl._create_unverified_context
(x_train, y_train), (x_test, y_test) = keras.datasets.mnist.load_data()
x_train = x_train.reshape([x_train.shape[0], -1])
x_test = x_test.reshape([x_test.shape[0], -1])
#导入 MNIST 数据集
model = keras.Sequential([
    layers.Dense(64, activation='relu', input_shape=(784,)),
    layers.BatchNormalization(),
    layers.Dense(64, activation='relu'),
    layers.BatchNormalization(),
    layers.Dense(64, activation='relu'),
    layers.BatchNormalization(),
    layers.Dense(10, activation='softmax')
```

```
])
#对每个层进行标准化,具体参数为 layers.BatchNormalization()
model.compile(optimizer=keras.optimizers.SGD(),
    loss=keras.losses.SparseCategoricalCrossentropy(),
    metrics=['accuracy'])
#编译模型
history = model.fit(x_train, y_train, batch_size=256, epochs=100, validation_split=0.3, verbose=0)
#制订训练计划,训练 100 次
plt.plot(history.history['accuracy'])
plt.plot(history.history['val_accuracy'])
plt.legend(['training', 'validation'], loc='upper left')
plt.show()
result = model.evaluate(x_test, y_test)
#打印输出结果
```

(4)批标准化的训练结果如图 4-4 所示。

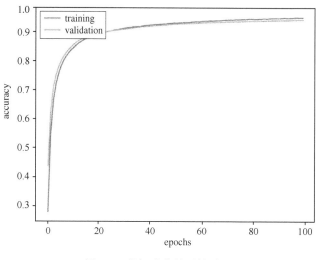

图 4-4　批标准化的训练结果

```
0000/10000 [==============================] - 0s 38us/sample - loss: 0.1807 - accuracy: 0.9463
```

本例采用了与基本数据集相同的数据集,并对每个层进行了批标准化。通过对比两者的图像和输出结果可以看出,在批标准化后,结果的精度和训练速度均有改善。

4.7　dropout

dropout(随机失活)是对具有深度结构的人工神经网络进行优化的方法,在学习过程中,通过将隐含层的部分权重或输出随机归零来降低节点间的依赖性、实现神经网络的正则化、降低其结构风险并实现优化目标。

（1）构建批标准化模型，输出图像和结果，代码如下。

```
#以 TensorFlow 为基础构建 Keras
import ssl
from tensorflow import keras
from tensorflow.keras import layers
#导入图形化工具
import matplotlib.pyplot as plt
#由于是 HTTPS 协议域名，需要添加相应证书
ssl._create_default_https_context = ssl._create_unverified_context
(x_train, y_train), (x_test, y_test) = keras.datasets.mnist.load_data()
x_train = x_train.reshape([x_train.shape[0], -1])
x_test = x_test.reshape([x_test.shape[0], -1])
#导入MNIST数据集
model = keras.Sequential([
    layers.Dense(64, activation='relu', input_shape=(784,)),
    layers.Dropout(0.2),
    layers.Dense(64, activation='relu'),
    layers.Dropout(0.2),
    layers.Dense(64, activation='relu'),
    layers.Dropout(0.2),
    layers.Dense(10, activation='softmax')
])
#对每个层的dropout进行设置，本例中采用的数值为0.2
model.compile(optimizer=keras.optimizers.SGD(),
    loss=keras.losses.SparseCategoricalCrossentropy(),
    metrics=['accuracy'])
#编译模型
history = model.fit(x_train, y_train, batch_size=256, epochs=100, validation_split=0.3, verbose=0)
#制订训练计划，训练100次
plt.plot(history.history['accuracy'])
plt.plot(history.history['val_accuracy'])
plt.legend(['training', 'validation'], loc='upper left')
plt.show()
result = model.evaluate(x_test, y_test)
#打印输出结果
```

（2）代码的运行结果如下，dropout 的训练结果如图 4-5 所示。

```
10000/10000 [==============================] - 0s 42us/sample - loss: 1.0328 - accuracy: 0.6482
```

本例采用了与基本数据集相同的数据集，并令每个层的 rate 为 0.2，以防止过拟合。其意义在于按比例 1/(1−rate)对层进行缩放，以使它们在训练时间和推理时间内的总和不变。

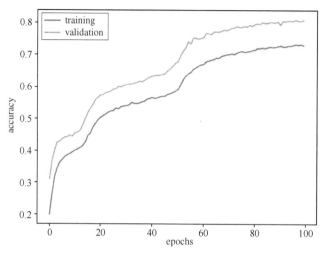

图 4-5 dropout 的训练结果

4.8 模型集成

在数据规模较大的情况下，单个模型的效率和预测结果并不能令人满意。因此，TensorFlow 2.0 提供了模型集成算法，可以将多个模型组合使用，能够得到更好的预测结果。

（1）构建批标准化模型，代码如下。

```
#以 TensorFlow 为基础构建 Keras
import numpy as np
from tensorflow import keras
from tensorflow.keras import layers
from tensorflow.keras.wrappers.scikit_learn import KerasClassifier
from sklearn.ensemble import VotingClassifier
from sklearn.metrics import accuracy_score
(x_train, y_train), (x_test, y_test) = keras.datasets.mnist.load_data()
x_train = x_train.reshape([x_train.shape[0], -1])
x_test = x_test.reshape([x_test.shape[0], -1])
#导入 MNIST 数据集
def mlp_model():
    model = keras.Sequential([
    layers.Dense(64, activation='relu', input_shape=(784,)),
    layers.Dropout(0.2),
    layers.Dense(64, activation='relu'),
    layers.Dropout(0.2),
    layers.Dense(64, activation='relu'),
    layers.Dropout(0.2),
    layers.Dense(10, activation='softmax')
    ])
    model.compile(optimizer=keras.optimizers.SGD(),
        loss=keras.losses.SparseCategoricalCrossentropy(),
```

```
        metrics=['accuracy'])
    return model
#制订一个 dropout 为 0.2 的基础模型
model1 = KerasClassifier(build_fn=mlp_model, epochs=100, verbose=0)
model2 = KerasClassifier(build_fn=mlp_model, epochs=100, verbose=0)
model3 = KerasClassifier(build_fn=mlp_model, epochs=100, verbose=0)
#使用 KerasClassifier，根据基础模型构建新的模型
ensemble_clf = VotingClassifier(estimators=[
    ('model1', model1), ('model2', model2), ('model3', model3)
], voting='soft')
#算法为 voting='soft'
ensemble_clf.fit(x_train, y_train)
y_pred = ensemble_clf.predict(x_test)
print('acc: ', accuracy_score(y_pred, y_test))
#输出结果
```

说明：voting='soft'表示本例使用投票法的软投票方式构建模型，即输出类概率。与之对应的还有 voting='hard'的硬投票方式，即输出类标签。

（2）代码的运行结果如下。

```
('acc: ', 0.9505)
```

本例使用了 accuracy_score，该函数会返回子集的准确率。如果一个样本必须严格匹配真实数据集中的 label，则整个集合的预测标签返回 1.0，否则返回 0.0。

4.9 优化器

优化器是一种扩展类，包含用于训练特定模型的附加信息。优化器使用给定的参数进行初始化，用于提高训练特定模型的速度和性能。根据实际情况选择相应的优化器是一种必要的优化手段。

（1）使用 SGD 优化器对模型进行优化，代码如下。

```
#以 TensorFlow 为基础构建 Keras
import ssl
from tensorflow import keras
from tensorflow.keras import layers
#导入图形化工具
import matplotlib.pyplot as plt
#由于是 HTTPS 协议域名，需要添加相应证书
ssl._create_default_https_context = ssl._create_unverified_context
(x_train, y_train), (x_test, y_test) = keras.datasets.mnist.load_data()
x_train = x_train.reshape([x_train.shape[0], -1])
x_test = x_test.reshape([x_test.shape[0], -1])
#导入 MNIST 数据集
model = keras.Sequential([
    layers.Dense(64, activation='sigmoid', input_shape=(784,)),
    layers.Dense(64, activation='sigmoid'),
```

```
        layers.Dense(64, activation='sigmoid'),
        layers.Dense(10, activation='softmax')
])
#构建模型
model.compile(optimizer=keras.optimizers.SGD(),
        loss=keras.losses.SparseCategoricalCrossentropy(),
        metrics=['accuracy'])
#编译模型,在编译过程中将 SGD 作为优化器
model.summary()
#输出各层结果
```

说明：SGD 优化器一般用来验证模型的收敛性,能够根据收敛速度调整学习速率,以使性能达到最优。

（2）代码的运行结果如下。

```
Model: "sequential"
_____
Layer (type)                 Output Shape              Param #
=================================================================
dense (Dense)                (None, 64)                50240
_____
dense_1 (Dense)              (None, 64)                4160
_____
dense_2 (Dense)              (None, 64)                4160
_____
dense_3 (Dense)              (None, 10)                650
=================================================================
Total params: 59,210
Trainable params: 59,210
Non-trainable params: 0
```

（3）训练 100 次后,显示图像,代码如下。

```
#以 TensorFlow 为基础构建 Keras
import ssl
from tensorflow import keras
from tensorflow.keras import layers
#导入图形化工具
import matplotlib.pyplot as plt
#由于是 HTTPS 协议域名,需要添加相应证书
ssl._create_default_https_context = ssl._create_unverified_context
(x_train, y_train), (x_test, y_test) = keras.datasets.mnist.load_data()
x_train = x_train.reshape([x_train.shape[0], -1])
x_test = x_test.reshape([x_test.shape[0], -1])
#导入 MNIST 数据集
model = keras.Sequential([
        layers.Dense(64, activation='sigmoid', input_shape=(784,)),
        layers.Dense(64, activation='sigmoid'),
```

```
        layers.Dense(64, activation='sigmoid'),
        layers.Dense(10, activation='softmax')
    ])
    #构建模型
    model.compile(optimizer=keras.optimizers.SGD(),
        loss=keras.losses.SparseCategoricalCrossentropy(),
        metrics=['accuracy'])
    #编译模型,在编译过程中将 SGD 作为优化器
    history = model.fit(x_train, y_train, batch_size=256, epochs=100, validation_split=
0.3, verbose=0)
    #制订训练计划,训练 100 次
    plt.plot(history.history['accuracy'])
    plt.plot(history.history['val_accuracy'])
    plt.legend(['training', 'validation'], loc='upper left')
    plt.show()
    result = model.evaluate(x_test, y_test)
    #打印输出结果
```

（4）代码的运行结果如下，优化器的训练结果如图 4-6 所示。

  0000/10000 [==============================] - 0s 40us/sample - loss: 2.4093 - accuracy: 0.1046

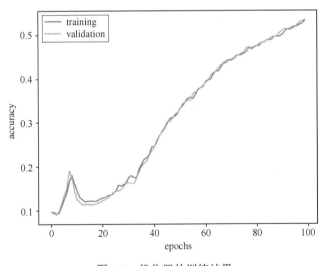

图 4-6 优化器的训练结果

  可以看出，在应用了优化器后，整体的训练结果得到优化。在实际的运行过程中，可以发现执行时间明显得到优化。

  本章介绍了 MLP 网络的基础知识，对前面章节中介绍的知识进行了简单的整理，能够加深读者对基础知识的认识。

第 5 章 TensorFlow 与卷积神经网络

卷积神经网络（Convolutional Neural Network，CNN）是一类包含卷积计算且具有深度结构的前馈神经网络（Feedforward Neural Network），是深度学习（Deep Learning）的代表算法之一。

5.1 基础卷积神经网络

卷积神经网络结构包括卷积层、降采样层、全连接层。本章通过几个例子简述基础 CNN 网络的构建。

在介绍基础 CNN 网络之前，需要构造数据集，这里采用经典的 MNIST 数据集。MNIST 数据集来自美国国家标准与技术研究所（National Institute of Standards and Technology，NIST）。训练集和测试集均由 250 个人手写的数字构成，其中 50%的人是高中生，其余 50%的人是来自美国人口普查局的工作人员。

（1）本例使用 MNIST 数据集构造数据集，代码如下。

```
import ssl
#引入 SSL 模块
from tensorflow import keras
from tensorflow.keras import layers
#引入 TensorFlow、Keras 模块
import matplotlib.pyplot as plt
#引入 matplotlib.pyplot 可视化模块
ssl._create_default_https_context = ssl._create_unverified_context
(x_train, y_train), (x_test, y_test) = keras.datasets.mnist.load_data()
#导入 MNIST 数据集
print(x_train.shape, ' ', y_train.shape)
print(x_test.shape, ' ', y_test.shape)
```

（2）运行程序，验证数据集的构造情况，结果如下。

```
(60000, 28, 28)   (60000,)
(10000, 28, 28)   (10000,)
```

（3）打印数据集中的图像，代码如下。

```
iimport ssl
#引入 SSL 模块
from tensorflow import keras
from tensorflow.keras import layers
#引入 TensorFlow、Keras 模块
import matplotlib.pyplot as plt
```

```
#引入matplotlib.pyplot 可视化模块
ssl._create_default_https_context = ssl._create_unverified_context
(x_train, y_train), (x_test, y_test) = keras.datasets.mnist.load_data()
#导入MNIST 数据集
plt.imshow(x_train[0])
plt.show()
#显示图像
```
（4）运行代码，得到MNIST数据集图像示例如图 5-1 所示。

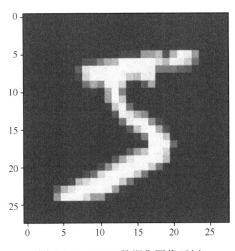

图 5-1　MNIST 数据集图像示例

5.2　卷积层的概念及示例

每个卷积层（Convolutional Layer）中都存在多个卷积单元，这些卷积单元的优化依赖反向传播算法。其目的是通过迭代输入提取更复杂的特征。

（1）本节使用构造的数据集构建卷积层，具体代码如下。

```
import ssl
#引入SSL 模块
from tensorflow import keras
from tensorflow.keras import layers
#引入TensorFlow、Keras 模块
import matplotlib.pyplot as plt
#引入matplotlib.pyplot 可视化模块
ssl._create_default_https_context = ssl._create_unverified_context
(x_train, y_train), (x_test, y_test) = keras.datasets.mnist.load_data()
#导入MNIST 数据集
x_train = x_train.reshape((-1,28,28,1))
x_test = x_test.reshape((-1,28,28,1))
model = keras.Sequential()
#使用序贯模型
```

```
model.add(layers.Conv2D(input_shape=(x_train.shape[0], x_train.shape[1],
    x_train.shape[2]), filters=32, kernel_size=(3,3),
    strides=(1,1), padding='valid', activation='relu'))
#使用 conv2d 构建卷积层
model.compile(optimizer=keras.optimizers.Adam(),
    loss=keras.losses.SparseCategoricalCrossentropy(),
    metrics=['accuracy'])
#编译模型
model.summary()
#输出各层的情况
```

(2) 代码的运行结果如下。

```
Model: " sequential "
_____
Layer (type)                 Output Shape              Param #
=================================================================
conv2d (Conv2D)              (None, 26, 26, 32)        320
=================================================================
Total params: 320
Trainable params: 320
Non-trainable params: 0
_____
```

本例构建了一个简单的卷积层，并输出了相应的层信息，卷积层是训练模型的基础。

5.3 池化层的概念及示例

池化（Pooling）是卷积神经网络中的重要概念，它通过降采样降低输入特征的维度。
(1) 在卷积层示例的基础上添加池化层，具体代码如下。

```
import ssl
#引入 SSL 模块
from tensorflow import keras
from tensorflow.keras import layers
#引入 TensorFlow、Keras 模块
import matplotlib.pyplot as plt
#引入 matplotlib.pyplot 可视化模块
ssl._create_default_https_context = ssl._create_unverified_context
(x_train, y_train), (x_test, y_test) = keras.datasets.mnist.load_data()
#导入 MNIST 数据集
x_train = x_train.reshape((-1,28,28,1))
x_test = x_test.reshape((-1,28,28,1))
model = keras.Sequential()
#使用序贯模型
model.add(layers.Conv2D(input_shape=(x_train.shape[0],x_train.shape[1],
```

```
        x_train.shape[2]), filters=32, kernel_size=(3,3),
        strides=(1,1), padding='valid', activation='relu'))
#使用 conv2d 构建卷积层
model.add(layers.MaxPool2D(pool_size=(2,2)))
#添加一个 pool_size=(2,2)的最大池化层
model.compile(optimizer=keras.optimizers.Adam(),
        loss=keras.losses.SparseCategoricalCrossentropy(),
        metrics=['accuracy'])
#编译模型
model.summary()
#输出各层的情况
```

说明：本例采用常用的池化方式，即每隔 2 个元素从图像中划分出 2×2 的区块，然后取每个区块中的 4 个数的最大值。

（2）代码的运行结果如下。

```
Model: " sequential "
_____
Layer (type)                 Output Shape              Param #
=================================================================
conv2d (Conv2D)              (None, 26, 26, 32)        320
_____
max_pooling2d (MaxPooling2D) (None, 13, 13, 32)        0
=================================================================
Total params: 320
Trainable params: 320
Non-trainable params: 0
_____
```

从结果中可以看出，已经添加了池化层。

5.4 全连接层的概念及示例

全连接层（Fully Connected Layers，FC）对卷积层和池化层学习到的特征进行分类，然后将这些特征映射到特定的样本标记空间。

（1）在池化层示例的基础上添加全连接层，具体代码如下。

```
import ssl
#引入 SSL 模块
from tensorflow import keras
from tensorflow.keras import layers
#引入 TensorFlow、Keras 模块
import matplotlib.pyplot as plt
#引入 matplotlib.pyplot 可视化模块
ssl._create_default_https_context = ssl._create_unverified_context
(x_train, y_train), (x_test, y_test) = keras.datasets.mnist.load_data()
```

```python
#导入MNIST数据集
x_train = x_train.reshape((-1,28,28,1))
x_test = x_test.reshape((-1,28,28,1))
model = keras.Sequential()
#使用序贯模型
model.add(layers.Conv2D(input_shape=(x_train.shape[0],x_train.shape[1],
    x_train.shape[2]), filters=32, kernel_size=(3,3),
    strides=(1,1),padding='valid',activation='relu'))
#使用conv2d构建卷积层
model.add(layers.MaxPool2D(pool_size=(2,2)))
#添加一个pool_size=(2,2)的最大池化层
model.add(layers.Flatten())
#将输入展平,不影响批量大小。这里使用默认值channels_last
model.add(layers.Dense(32, activation='relu'))
#分类层
model.add(layers.Dense(10, activation='softmax'))
#连接层,相当于添加一个层
model.compile(optimizer=keras.optimizers.Adam(),
    loss=keras.losses.SparseCategoricalCrossentropy(),
    metrics=['accuracy'])
#编译模型
model.summary()
#输出各层的情况
```

(2)代码的运行结果如下。

```
Model: "sequential"
_____
Layer (type)                 Output Shape              Param #
=================================================================
conv2d (Conv2D)              (None, 26, 26, 32)        320
_____
max_pooling2d (MaxPooling2D) (None, 13, 13, 32)        0
_____
flatten (Flatten)            (None, 5408)              0
_____
dense (Dense)                (None, 32)                173088
_____
dense_1 (Dense)              (None, 10)                330
=================================================================
Total params: 173,738
Trainable params: 173,738
Non-trainable params: 0
_____
```

本节构建了全连接层,并使用了默认的参数执行代码。在结果中显示了全连接层的情况。

5.5 模型的概念、配置及训练

模型的本质是函数,具有函数的特性,即对输入的数据进行变换后,输出相应的数据。一个问题可以构建出多个模型,选择"合适"的模型是机器学习的重点。

(1) 本节使用 5.4 节构建的添加全连接层的例子进行介绍,具体代码如下。

```
import ssl
#引入 SSL 模块
from tensorflow import keras
from tensorflow.keras import layers
#引入 TensorFlow、Keras 模块
import matplotlib.pyplot as plt
#引入 matplotlib.pyplot 可视化模块
ssl._create_default_https_context = ssl._create_unverified_context
(x_train, y_train), (x_test, y_test) = keras.datasets.mnist.load_data()
#导入 MNIST 数据集
model = keras.Sequential()
#使用序贯模型
model.add(layers.Conv2D(input_shape=(x_train.shape[0], x_train.shape[1],
    x_train.shape[2]), filters=32, kernel_size=(3,3),
    strides=(1,1),padding='valid', activation='relu'))
#使用 conv2d 构建卷积层
model.add(layers.MaxPool2D(pool_size=(2,2)))
#添加一个 pool_size=(2,2)的最大池化层
model.add(layers.Flatten())
#将输入展平,不影响批量大小。这里使用默认值 channels_last
model.add(layers.Dense(32, activation='relu'))
#分类层
model.add(layers.Dense(10, activation='softmax'))
#连接层,相当于添加一个层
model.compile(optimizer=keras.optimizers.Adam(),
    loss=keras.losses.SparseCategoricalCrossentropy(),
    metrics=['accuracy'])
#编译模型
history = model.fit(x_train, y_train, batch_size=64, epochs=5, validation_split=0.1)
#训练 5 次
```

(2) 代码的运行结果如下。

```
Train on 54000 samples, validate on 6000 samples
Epoch 1/5
54000/54000 [==============================] - 14s 268us/sample - loss: 1.2469 -
```

```
accuracy: 0.6450 - val_loss: 0.5117 - val_accuracy: 0.8953
...
Epoch 5/5
54000/54000 [==============================] - 13s 244us/sample - loss: 0.0686 - accuracy: 0.9787 - val_loss: 0.1121 - val_accuracy: 0.9718
```

（3）在多次训练后显示图像，训练结果如图 5-2 所示。

```
import ssl
#引用 SSL 模块
from tensorflow import keras
from tensorflow.keras import layers
#引入 TensorFlow、Keras 模块
import matplotlib.pyplot as plt
#引入 matplotlib.pyplot 可视化模块
ssl._create_default_https_context = ssl._create_unverified_context
(x_train, y_train), (x_test, y_test) = keras.datasets.mnist.load_data()
#导入 MNIST 数据集
model = keras.Sequential()
#使用序贯模型
model.add(layers.Conv2D(input_shape=(x_train.shape[0],x_train.shape[1],
    x_train.shape[2]),filters=32, kernel_size=(3,3),
    strides=(1,1),padding='valid', activation='relu'))
#使用 conv2d 构建卷积层
model.add(layers.MaxPool2D(pool_size=(2,2)))
#添加一个 pool_size=(2,2)的最大池化层
model.add(layers.Flatten())
#将输入展平，不影响批量大小。这里使用默认值 channels_last
model.add(layers.Dense(32, activation='relu'))
#分类层
model.add(layers.Dense(10, activation='softmax'))
#连接层，相当于添加一个层
model.compile(optimizer=keras.optimizers.Adam(),
    loss=keras.losses.SparseCategoricalCrossentropy(),
    metrics=['accuracy'])
#编译创建的模型
history = model.fit(x_train, y_train, batch_size=64, epochs=5, validation_split=0.1)
#训练 5 次
plt.plot(history.history['accuracy'])
plt.plot(history.history['val_accuracy'])
plt.legend(['training', 'valivation'], loc='upper left')
plt.show()
#显示训练结果
```

第 5 章 TensorFlow 与卷积神经网络

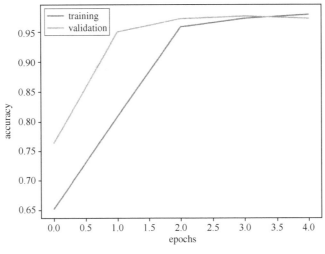

图 5-2 训练结果

第6章 TensorFlow 自编码器

自编码器将输入信息作为学习目标，并对其进行表征学习。

6.1 自编码器简介

自编码器包含编码器和解码器。编码器提供按规则编码的功能，解码器将编码器的输出扩展为与编码器输入具有相同维度的输出。在这个过程中，自编码器通过数据重组进行机器学习。

（1）本节使用 MNIST 数据集对自编码器进行介绍，代码如下。

```
#以 TensorFlow 为基础构建 Keras
import tensorflow as tf
import tensorflow.keras as keras
import tensorflow.keras.layers as layers
from IPython.display import SVG
#打印 TensorFlow 版本
print(tf.__version__)
(x_train, y_train), (x_test, y_test) = keras.datasets.mnist.load_data()
x_train = x_train.reshape((-1, 28*28)) / 255.0
x_test = x_test.reshape((-1, 28*28)) / 255.0
print(x_train.shape, ' ', y_train.shape)
print(x_test.shape, ' ', y_test.shape)
```

（2）代码的运行结果如下。

```
2.0.0-rc2
(60000, 784)    (60000,)
(10000, 784)    (10000,)
```

（3）构建一个简单的自编码器，代码如下。

```
#以 TensorFlow 为基础构建 Keras
import tensorflow as tf
import tensorflow.keras as keras
import tensorflow.keras.layers as layers
from IPython.display import SVG
#打印 TensorFlow 版本
print(tf.__version__)
#导入数据集并划分训练集和测试集
(x_train, y_train), (x_test, y_test) = keras.datasets.mnist.load_data()
x_train = x_train.reshape((-1, 28*28)) / 255.0
x_test = x_test.reshape((-1, 28*28)) / 255.0
```

```
code_dim = 32
#建立输入层
inputs = layers.Input(shape=(x_train.shape[1],), name='inputs')
code = layers.Dense(code_dim, activation='relu', name='code')(inputs)
outputs = layers.Dense(x_train.shape[1], activation='softmax', name='outputs')(code)
#构建自编码器
auto_encoder = keras.Model(inputs, outputs)
auto_encoder.summary()
```

（4）代码的运行结果如下。

```
Model: " model "
_____
Layer (type)                 Output Shape              Param #
=================================================================
inputs (InputLayer)          [(None, 784)]             0
_____
code (Dense)                 (None, 32)                25120
_____
outputs (Dense)              (None, 784)               25872
=================================================================
Total params: 50,992
Trainable params: 50,992
Non-trainable params: 0
_____
```

（5）编码后需要进行解码操作，训练10次，代码如下。

```
#以 TensorFlow 为基础构建 Keras
import tensorflow as tf
import tensorflow.keras as keras
import tensorflow.keras.layers as layers
from IPython.display import SVG
#打印 TensorFlow 版本
print(tf.__version__)
#导入数据集并划分训练集和测试集
(x_train, y_train), (x_test, y_test) = keras.datasets.mnist.load_data()
x_train = x_train.reshape((-1, 28*28)) / 255.0
x_test = x_test.reshape((-1, 28*28)) / 255.0
code_dim = 32
#建立输入层
inputs = layers.Input(shape=(x_train.shape[1],), name='inputs')
code = layers.Dense(code_dim, activation='relu', name='code')(inputs)
outputs = layers.Dense(x_train.shape[1], activation='softmax', name='outputs')(code)
#构建自编码器
```

```
auto_encoder = keras.Model(inputs, outputs)
#使用自编码器对设置的层进行编码
keras.utils.plot_model(auto_encoder, show_shapes=True)
encoder = keras.Model(inputs,code)
keras.utils.plot_model(encoder, show_shapes=True)
#构建解码器
decoder_input = keras.Input((code_dim,))
decoder_output = auto_encoder.layers[-1](decoder_input)
decoder = keras.Model(decoder_input, decoder_output)
keras.utils.plot_model(decoder, show_shapes=True)
#使用解码器对编码后的层解码
auto_encoder.compile(optimizer='adam', loss='binary_crossentropy')
#对设置的模型进行训练
history = auto_encoder.fit(x_train, x_train, batch_size=64, epochs=10, validation_split=0.1)
```

（6）代码的运行结果如下。

```
Train on 54000 samples, validate on 6000 samples
Epoch 1/10
Instructions for updating:
Use tf.where in 2.0, which has the same broadcast rule as np.where
54000/54000 [==============================] - 5s 92us/sample - loss: 0.7058 - val_loss: 0.6797
...
Epoch 10/10
54000/54000 [==============================] - 4s 80us/sample - loss: 0.6740 - val_loss: 0.6710
```

（7）对训练模型的结果进行验证，代码如下。

```
#以 TensorFlow 为基础构建 Keras
import tensorflow as tf
import tensorflow.keras as keras
import tensorflow.keras.layers as layers
from IPython.display import SVG
#打印 TensorFlow 版本
print(tf.__version__)
#导入数据集并划分训练集和测试集
(x_train, y_train), (x_test, y_test) = keras.datasets.mnist.load_data()
x_train = x_train.reshape((-1, 28*28)) / 255.0
x_test = x_test.reshape((-1, 28*28)) / 255.0
code_dim = 32
#建立输入层
inputs = layers.Input(shape=(x_train.shape[1],), name='inputs')
code = layers.Dense(code_dim, activation='relu', name='code')(inputs)
outputs = layers.Dense(x_train.shape[1], activation='softmax', name='outputs')(code)
```

```python
#构建自编码器
auto_encoder = keras.Model(inputs, outputs)
#使用自编码器对设置的层进行编码
keras.utils.plot_model(auto_encoder, show_shapes=True)
encoder = keras.Model(inputs,code)
keras.utils.plot_model(encoder, show_shapes=True)
#构建解码器
decoder_input = keras.Input((code_dim,))
decoder_output = auto_encoder.layers[-1](decoder_input)
decoder = keras.Model(decoder_input, decoder_output)
keras.utils.plot_model(decoder, show_shapes=True)
#使用解码器对编码后的层进行解码
auto_encoder.compile(optimizer='adam', loss='binary_crossentropy')
#对设置的模型进行训练
history = auto_encoder.fit(x_train, x_train, batch_size=64, epochs=10, validation_split=0.1)
#对自编码器和解码器进行预测
encoded = encoder.predict(x_test)
decoded = decoder.predict(encoded)
import matplotlib.pyplot as plt
plt.figure(figsize=(10,4))
#循环构建数据并显示图像
n = 5
for i in range(n):
    ax = plt.subplot(2, n, i+1)
    plt.imshow(x_test[i].reshape(28,28))
    plt.gray()
    ax.get_xaxis().set_visible(False)
    ax.get_yaxis().set_visible(False)
    ax = plt.subplot(2, n, n+i+1)
    plt.imshow(decoded[i].reshape(28,28))
    plt.gray()
    ax.get_xaxis().set_visible(False)
    ax.get_yaxis().set_visible(False)
plt.show()
```

（8）自编码器的运行结果如图 6-1 所示。

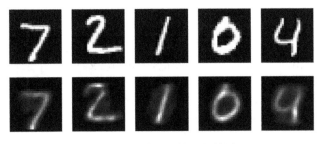

图 6-1　自编码器的运行结果

6.2 卷积自编码器

与基础自编码器不同，卷积自编码器用卷积层代替了全连接层，目的是降低输入特征的维度，使速度更快、准确率更高。

（1）本节使用 MNIST 数据集对卷积自编码器进行介绍，代码如下。

```
#以 TensorFlow 为基础构建 Keras
import tensorflow as tf
import tensorflow.keras as keras
import tensorflow.keras.layers as layers
#打印 TensorFlow 版本
print(tf.__version__)
#导入数据集并划分训练集和测试集
(x_train, y_train), (x_test, y_test) = keras.datasets.mnist.load_data()
x_train = tf.expand_dims(x_train.astype('float32'), -1) / 255.0
x_test = tf.expand_dims(x_test.astype('float32'),-1) / 255.0
print(x_train.shape, ' ', y_train.shape)
print(x_test.shape, ' ', y_test.shape)
```

（2）代码的运行结果如下。

```
2.0.0-rc2
(60000, 28, 28, 1)    (60000,)
(10000, 28, 28, 1)    (10000,)
```

（3）验证数据，代码如下。

```
#以 TensorFlow 为基础构建 Keras
import tensorflow as tf
import tensorflow.keras as keras
import tensorflow.keras.layers as layers
import pydot
import pydot_ng as pydot
#导入数据集并划分训练集和测试集
(x_train, y_train), (x_test, y_test) = keras.datasets.mnist.load_data()
x_train = tf.expand_dims(x_train.astype('float32'), -1) / 255.0
x_test = tf.expand_dims(x_test.astype('float32'),-1) / 255.0
#建立输入层
inputs = layers.Input(shape=(x_train.shape[1], x_train.shape[2], x_train.shape[3]), name='inputs')
print(inputs.shape)
code = layers.Conv2D(16, (3,3), activation='relu', padding='same')(inputs)
code = layers.MaxPool2D((2,2), padding='same')(code)
print(code.shape)
decoded = layers.Conv2D(16, (3,3), activation='relu', padding='same')(code)
decoded = layers.UpSampling2D((2,2))(decoded)
print(decoded.shape)
```

```
#建立输出层
outputs = layers.Conv2D(1, (3,3), activation='sigmoid', padding='same')(decoded)
print(outputs.shape)
```

(4) 代码的运行结果如下。

```
(None, 28, 28, 1)
(None, 14, 14, 16)
(None, 28, 28, 16)
(None, 28, 28, 1)
```

(5) 构建模型并训练,代码如下。

```
#以 TensorFlow 为基础构建 Keras
import tensorflow as tf
import tensorflow.keras as keras
import tensorflow.keras.layers as layers
import pydot
import pydot_ng as pydot
#导入数据集并划分训练集和测试集
(x_train, y_train), (x_test, y_test) = keras.datasets.mnist.load_data()
x_train = tf.expand_dims(x_train.astype('float32'), -1) / 255.0
x_test = tf.expand_dims(x_test.astype('float32'),-1) / 255.0
#建立输入层
inputs = layers.Input(shape=(x_train.shape[1], x_train.shape[2], x_train.shape[3]), name='inputs')
code = layers.Conv2D(16, (3,3), activation='relu', padding='same')(inputs)
code = layers.MaxPool2D((2,2), padding='same')(code)
decoded = layers.Conv2D(16, (3,3), activation='relu', padding='same')(code)
decoded = layers.UpSampling2D((2,2))(decoded)
#建立输出层
outputs = layers.Conv2D(1, (3,3), activation='sigmoid', padding='same')(decoded)
#构建模型
auto_encoder = keras.Model(inputs, outputs)
auto_encoder.compile(optimizer=keras.optimizers.Adam(),
    loss=keras.losses.BinaryCrossentropy())
keras.utils.plot_model(auto_encoder, show_shapes=True)
#定义模型训练的回调函数,达到一定数值后停止训练
early_stop = keras.callbacks.EarlyStopping(patience=2, monitor='loss')
auto_encoder.fit(x_train,x_train, batch_size=64, epochs=1, validation_split=0.1, validation_freq=10, callbacks=[early_stop])
```

(6) 代码的运行结果如下。

```
Train on 54000 samples, validate on 6000 samples
54000/54000 [==============================] - 166s 3ms/sample - loss: 0.1077
```

(7) 对模型进行训练并测试,显示图像,代码如下。

```
#以 TensorFlow 为基础构建 Keras
import tensorflow as tf
import tensorflow.keras as keras
```

```python
import tensorflow.keras.layers as layers
import pydot
import pydot_ng as pydot
#导入数据集并划分训练集和测试集
(x_train, y_train), (x_test, y_test) = keras.datasets.mnist.load_data()
x_train = tf.expand_dims(x_train.astype('float32'), -1) / 255.0
x_test = tf.expand_dims(x_test.astype('float32'),-1) / 255.0
#建立输入层
inputs = layers.Input(shape=(x_train.shape[1], x_train.shape[2], x_train.shape[3]), name='inputs')
code = layers.Conv2D(16, (3,3), activation='relu', padding='same')(inputs)
code = layers.MaxPool2D((2,2), padding='same')(code)
decoded = layers.Conv2D(16, (3,3), activation='relu', padding='same')(code)
decoded = layers.UpSampling2D((2,2))(decoded)
#建立输出层
outputs = layers.Conv2D(1, (3,3), activation='sigmoid', padding='same')(decoded)
#构建模型
auto_encoder = keras.Model(inputs, outputs)
auto_encoder.compile(optimizer=keras.optimizers.Adam(),
    loss=keras.losses.BinaryCrossentropy())
keras.utils.plot_model(auto_encoder, show_shapes=True)
#定义模型训练的回调函数,达到一定数值后停止训练
early_stop = keras.callbacks.EarlyStopping(patience=2, monitor='loss')
auto_encoder.fit(x_train,x_train, batch_size=64, epochs=1, validation_split=0.1,
        validation_freq=10, callbacks=[early_stop])
#显示图像
import matplotlib.pyplot as plt
decoded = auto_encoder.predict(x_test)
n = 5
plt.figure(figsize=(10, 4))
for i in range(n):
    # display original
    ax = plt.subplot(2, n, i+1)
    plt.imshow(tf.reshape(x_test[i+1],(28, 28)))
    plt.gray()
    ax.get_xaxis().set_visible(False)
    ax.get_yaxis().set_visible(False)
    ax = plt.subplot(2, n, i + n+1)
    plt.imshow(tf.reshape(decoded[i+1],(28, 28)))
    plt.gray()
    ax.get_xaxis().set_visible(False)
    ax.get_yaxis().set_visible(False)
plt.show()
```

(8) 卷积自编码器的运行结果如图 6-2 所示。

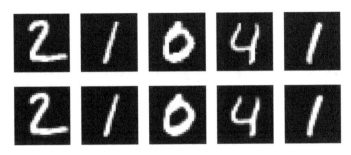

图 6-2　卷积自编码器的运行结果

第 7 章 TensorFlow 高级编程

与 TensorFlow 1.x 版本相比，TensorFlow 2.0 做了较大的更改，使用大量 Keras 作为默认高级 API，大大降低了 TensorFlow 的使用难度。本章对这些 API 进行简单的介绍。

7.1 Keras 基础

Keras 是一个由 Python 编写的开源人工神经网络库，可以作为 TensorFlow、CNTK 和 Theano 的高阶应用程序接口，实现深度学习模型的设计、调试、评估、应用和可视化。

Keras 有 3 个优点：方便用户使用、模块化和可组合、易于扩展。TensorFlow 2.0 推荐使用 Keras 构建网络，常见的神经网络都包含在 keras.layer 中。

（1）本例使用如下代码进行测试。

```
#以 TensorFlow 为基础构建 Keras
import tensorflow as tf
from tensorflow.keras import layers
#打印 TensorFlow 版本
print(tf.__version__)
#打印 Keras 版本
print(tf.keras.__version__)
```

（2）代码的运行结果如下。

```
2.0.0-alpha0
2.2.4-tf
```

7.1.1 构造数据

所有机器学习都以数据为基础，本节介绍如何使用 Keras 构造数据。

（1）Keras 构造数据的基础函数是 tf.data，本例使用 tf.data 构造数据，代码如下。

```
#以 TensorFlow 为基础构建 Keras
from __future__ import absolute_import, division, print_function
import tensorflow as tf
from tensorflow import keras
from tensorflow.keras import layers
#导入数据集并划分训练集和测试集
(x_train, y_train), (x_test, y_test) = keras.datasets.mnist.load_data()
x_train = x_train.reshape(60000, 784).astype('float32') /255
x_test = x_test.reshape(10000, 784).astype('float32') /255
#以 10000 划分数据集
```

```python
x_val = x_train[-10000:]
y_val = y_train[-10000:]
x_train = x_train[:-10000]
y_train = y_train[:-10000]
#构建模型函数
def get_compiled_model():
    #建立输入层
    inputs = keras.Input(shape=(784,), name='mnist_input')
    h1 = layers.Dense(64, activation='relu')(inputs)
    h2 = layers.Dense(64, activation='relu')(h1)
    #建立输出层
    outputs = layers.Dense(10, activation='softmax')(h2)
    #使用 Keras 构建模型
    model = keras.Model(inputs, outputs)
    model.compile(optimizer=keras.optimizers.RMSprop(),
        loss=keras.losses.SparseCategoricalCrossentropy(),
        metrics=[keras.metrics.SparseCategoricalAccuracy()])
    return model
    #定义模型
model = get_compiled_model()
#建立训练集
train_dataset = tf.data.Dataset.from_tensor_slices((x_train, y_train))
train_dataset = train_dataset.shuffle(buffer_size=1024).batch(64)
#建立测试集
val_dataset = tf.data.Dataset.from_tensor_slices((x_val, y_val))
val_dataset = val_dataset.batch(64)
#model.fit(train_dataset, epochs=3)
#steps_per_epoch 每个 epoch 只训练几步
#validation_steps 每次验证几步
model.fit(train_dataset, epochs=3, steps_per_epoch=100,
    validation_data=val_dataset, validation_steps=3)
```

(2)代码的运行结果如下。

```
Train for 100 steps, validate for 3 steps
Epoch 1/3
100/100 [==============================] - 2s 17ms/step - loss: 0.7462 - sparse_categorical_accuracy: 0.8016 - val_loss: 0.4028 - val_sparse_categorical_accuracy: 0.8802
Epoch 2/3
100/100 [==============================] - 0s 4ms/step - loss: 0.3714 - sparse_categorical_accuracy: 0.8963 - val_loss: 0.3114 - val_sparse_categorical_accuracy: 0.8802
Epoch 3/3
100/100 [==============================] - 0s 4ms/step - loss: 0.3268 - sparse_categorical_accuracy: 0.9023 - val_loss: 0.2508 - val_sparse_categorical_accuracy: 0.9167
```

7.1.2 样本权重和类权重

样本权重数组用于指定批处理中每个样本在计算损失率时应具有的权重值,通常用于处理不平衡的分类问题。当使用的权重是 1 和 0 时,该数组可以作为损失函数的掩码。

类权重更加具体,它将类索引映射到应该用于属于该类的样本的样本权重。

(1) 本节对样本权重和类权重进行介绍,代码如下。

```
#以 TensorFlow 为基础构建 Keras
from __future__ import absolute_import, division, print_function
import tensorflow as tf
from tensorflow import keras
from tensorflow.keras import layers
#导入数据集并划分训练集和测试集
(x_train, y_train), (x_test, y_test) = keras.datasets.mnist.load_data()
x_train = x_train.reshape(60000, 784).astype('float32') /255
x_test = x_test.reshape(10000, 784).astype('float32') /255
#以 10000 划分数据集
x_val = x_train[-10000:]
y_val = y_train[-10000:]
x_train = x_train[:-10000]
y_train = y_train[:-10000]
#构建模型函数
def get_compiled_model():
#建立输入层
    inputs = keras.Input(shape=(784,), name='mnist_input')
    h1 = layers.Dense(64, activation='relu')(inputs)
    h2 = layers.Dense(64, activation='relu')(h1)
#建立输出层
    outputs = layers.Dense(10, activation='softmax')(h2)
#使用 Keras 构建模型
    model = keras.Model(inputs, outputs)
    model.compile(optimizer=keras.optimizers.RMSprop(),
        loss=keras.losses.SparseCategoricalCrossentropy(),
        metrics=[keras.metrics.SparseCategoricalAccuracy()])
    return model
#增加第 5 类的权重
import numpy as np
#类权重
model = get_compiled_model()
class_weight = {i:1.0 for i in range(10)}
class_weight[5] = 2.0
print(class_weight)
model.fit(x_train, y_train, class_weight=class_weight, batch_size=64, epochs=4)
```

#样本权重
```
model = get_compiled_model()
sample_weight = np.ones(shape=(len(y_train),))
sample_weight[y_train == 5] = 2.0
model.fit(x_train, y_train, sample_weight=sample_weight, batch_size=64, epochs=4)
#tf.data 数据
model = get_compiled_model()
sample_weight = np.ones(shape=(len(y_train),))
sample_weight[y_train == 5] = 2.0
train_dataset = tf.data.Dataset.from_tensor_slices((x_train, y_train, sample_weight))
train_dataset = train_dataset.shuffle(buffer_size=1024).batch(64)
val_dataset = tf.data.Dataset.from_tensor_slices((x_val, y_val))
val_dataset = val_dataset.batch(64)
#训练 3 次
model.fit(train_dataset, epochs=3, )
```

（2）代码的运行结果如下。

```
{0: 1.0, 1: 1.0, 2: 1.0, 3: 1.0, 4: 1.0, 5: 2.0, 6: 1.0, 7: 1.0, 8: 1.0, 9: 1.0}
Train on 50000 samples
Epoch 1/4
50000/50000 [==============================] - 4s 71us/sample - loss: 0.3767 - sparse_categorical_accuracy: 0.9013
...
Epoch 4/4
50000/50000 [==============================] - 3s 52us/sample - loss: 0.0997 - sparse_categorical_accuracy: 0.9714
Train on 50000 samples
Epoch 1/4
50000/50000 [==============================] - 3s 62us/sample - loss: 0.3806 - sparse_categorical_accuracy: 0.8982
...
Epoch 4/4
50000/50000 [==============================] - 2s 50us/sample - loss: 0.0974 - sparse_categorical_accuracy: 0.9725
Epoch 1/3
782/Unknown - 4s 5ms/step - loss: 0.3741 - sparse_categorical_accuracy: 0.90252019- 10-09 11:51:43.355055:    W    tensorflow/core/common_runtime/base_collective_executor.    cc:216] BaseCollectiveExecutor::StartAbort Out of range: End of sequence
     [[{{node IteratorGetNext}}]]
782/782 [==============================] - 4s 5ms/step - loss: 0.3741 - sparse_categorical_accuracy: 0.9025
...
Epoch 3/3
```

```
781/782 [============================>.] - ETA: 0s - loss: 0.1358 - sparse_
categorical_accuracy: 0.96222019-10-09 11:51:49.115580: W tensorflow/core/common_runtime/base_
collective_executor.cc:216] BaseCollectiveExecutor::StartAbort Out of range: End of sequence
         [[{{node IteratorGetNext}}]]
782/782 [==============================] - 3s 4ms/step - loss: 0.1358 - sparse_
categorical_accuracy: 0.9622
```

7.1.3 回调

Keras 中的回调是在训练期间（epoch 开始时、batch 结束时、epoch 结束时等）不同点处调用的对象。

（1）本节对回调进行介绍，代码如下。

```
#以 TensorFlow 为基础构建 Keras
from __future__ import absolute_import, division, print_function
import tensorflow as tf
from tensorflow import keras
from tensorflow.keras import layers
#导入数据集并划分训练集和测试集
(x_train, y_train), (x_test, y_test) = keras.datasets.mnist.load_data()
x_train = x_train.reshape(60000, 784).astype('float32') /255
x_test = x_test.reshape(10000, 784).astype('float32') /255
#以 10000 划分数据集
x_val = x_train[-10000:]
y_val = y_train[-10000:]
x_train = x_train[:-10000]
y_train = y_train[:-10000]
#构建模型函数
def get_compiled_model():
#建立输入层
    inputs = keras.Input(shape=(784,), name='mnist_input')
    h1 = layers.Dense(64, activation='relu')(inputs)
    h2 = layers.Dense(64, activation='relu')(h1)
    #建立输出层
    outputs = layers.Dense(10, activation='softmax')(h2)
    #使用 Keras 构建模型
    model = keras.Model(inputs, outputs)
    model.compile(optimizer=keras.optimizers.RMSprop(),
        loss=keras.losses.SparseCategoricalCrossentropy(),
        metrics=[keras.metrics.SparseCategoricalAccuracy()])
    return model
#使用自定义模型
model = get_compiled_model()
#建立回调函数
```

第7章 TensorFlow 高级编程

```
callbacks = [
    keras.callbacks.EarlyStopping(
        #是否有提升关注的指标
        monitor='val_loss',
        #不再提升的阈值
        min_delta=1e-2,
        #2 个 epoch 没有提升就停止
        patience=2,
        verbose=1)
]
#训练模型
model.fit(x_train, y_train, epochs=20, batch_size=64,
    callbacks=callbacks, validation_split=0.2)
#checkpoint 模型回调
model = get_compiled_model()
check_callback = keras.callbacks.ModelCheckpoint(
    filepath='mymodel_{epoch}.h5',
    save_best_only=True,
    monitor='val_loss',
    verbose=1
)
#训练模型
model.fit(x_train, y_train, epochs=3, batch_size=64,
    callbacks=[check_callback], validation_split=0.2)
#动态调整学习率
initial_learning_rate = 0.1
lr_schedule = keras.optimizers.schedules.ExponentialDecay(
    initial_learning_rate,
    decay_steps=10000,
    decay_rate=0.96,
    staircase=True
)
optimizer = keras.optimizers.RMSprop(learning_rate=lr_schedule)
#使用 tensorboard
tensorboard_cbk = keras.callbacks.TensorBoard(log_dir='./7_1_test_log')
model.fit(x_train, y_train, epochs=5, batch_size=64,
    callbacks=[tensorboard_cbk], validation_split=0.2)
```

说明：tensorboard 是 TensorFlow 自带的可视化学习组件。

（2）代码的运行结果如下。

```
Train on 40000 samples, validate on 10000 samples
Epoch 1/20
40000/40000 [==============================] - 3s 66us/sample - loss: 0.3710 - sparse_categorical_accuracy: 0.8947 - val_loss: 0.2337 - val_sparse_categorical_accuracy: 0.9285
```

```
    ...
    Epoch 7/20
    40000/40000 [==============================] - 2s 46us/sample - loss: 0.0602 - sparse_
categorical_accuracy: 0.9818 - val_loss: 0.1355 - val_sparse_categorical_accuracy: 0.9616
    Epoch 00007: early stopping
    Train on 40000 samples, validate on 10000 samples
    Epoch 1/3
    39552/40000 [=============================>.] - ETA: 0s - loss: 0.3826 - sparse_
categorical_accuracy: 0.8939
    Epoch 00001: val_loss improved from inf to 0.24057, saving model to mymodel_1.h5
    40000/40000 [==============================] - 3s 79us/sample - loss: 0.3808 - sparse_
categorical_accuracy: 0.8943 - val_loss: 0.2406 - val_sparse_categorical_accuracy: 0.9293
    ....
    Epoch 3/3
    39296/40000 [=============================>.] - ETA: 0s - loss: 0.1364 - sparse_
categorical_accuracy: 0.9597
    Epoch 00003: val_loss improved from 0.18990 to 0.17021, saving model to mymodel_3.h5
    40000/40000 [==============================] - 2s 57us/sample - loss: 0.1356 - sparse_
categorical_accuracy: 0.9600 - val_loss: 0.1702 - val_sparse_categorical_accuracy: 0.9484
    Train on 40000 samples, validate on 10000 samples
    Epoch 1/5
    2019-10-09 11:59:54.702064: I tensorflow/core/profiler/lib/profiler_session.cc:184]
Profiler session started.
    40000/40000 [==============================] - 2s 52us/sample - loss: 0.1075 - sparse_
categorical_accuracy: 0.9675 - val_loss: 0.1481 - val_sparse_categorical_accuracy: 0.9557
    ...
    Epoch 5/5
    40000/40000 [==============================] - 2s 49us/sample - loss: 0.0546 - sparse_
categorical_accuracy: 0.9840 - val_loss: 0.1585 - val_sparse_categorical_accuracy: 0.9590
```

（3）当常用的回调方法不能满足实际使用需求的情况下，可以按照实际情况自定义回调方法，示例代码如下。

```
#以 TensorFlow 为基础构建 Keras
from __future__ import absolute_import, division, print_function
import tensorflow as tf
from tensorflow import keras
from tensorflow.keras import layers
#导入数据集并划分训练集和测试集
(x_train, y_train), (x_test, y_test) = keras.datasets.mnist.load_data()
x_train = x_train.reshape(60000, 784).astype('float32') /255
x_test = x_test.reshape(10000, 784).astype('float32') /255
#以 10000 划分数据集
x_val = x_train[-10000:]
y_val = y_train[-10000:]
```

```
    x_train = x_train[:-10000]
    y_train = y_train[:-10000]
    #构建模型函数
    def get_compiled_model():
    #建立输入层
        inputs = keras.Input(shape=(784,), name='mnist_input')
        h1 = layers.Dense(64, activation='relu')(inputs)
        h2 = layers.Dense(64, activation='relu')(h1)
        #建立输出层
        outputs = layers.Dense(10, activation='softmax')(h2)
        #使用 Keras 构建模型
        model = keras.Model(inputs, outputs)
        model.compile(optimizer=keras.optimizers.RMSprop(),
            loss=keras.losses.SparseCategoricalCrossentropy(),
            metrics=[keras.metrics.SparseCategoricalAccuracy()])
        return model
    #自定义方法,以 loss 为例
    class LossHistory(keras.callbacks.Callback):
        def on_train_begin(self, logs):
            self.losses = []
        def on_epoch_end(self, batch, logs):
            self.losses.append(logs.get('loss'))
            print('\nloss:',self.losses[-1])
    #使用自定义模型
    model = get_compiled_model()
    #使用自定义方法构建回调
    callbacks = [
        LossHistory()
    ]
    #对模型进行训练
    model.fit(x_train, y_train, epochs=3, batch_size=64,
        callbacks=callbacks, validation_split=0.2)
```

(4)代码的运行结果如下。

```
Train on 40000 samples, validate on 10000 samples
Epoch 1/3
    39360/40000 [============================>.] - ETA: 0s - loss: 0.3576 - sparse_categorical_accuracy: 0.8988
    loss: 0.3558141928195953
    40000/40000 [==============================] - 3s 67us/sample - loss: 0.3558 - sparse_categorical_accuracy: 0.8992 - val_loss: 0.2152 - val_sparse_categorical_accuracy: 0.9341
    ...
    Epoch 3/3
    39616/40000 [============================>.] - ETA: 0s - loss: 0.1225 - sparse_
```

```
categorical_accuracy: 0.9638
       loss: 0.12251021726131439
       40000/40000 [==============================] - 2s 46us/sample - loss: 0.1225 - sparse_
categorical_accuracy: 0.9639 - val_loss: 0.1904 - val_sparse_categorical_accuracy: 0.9433
```

本节介绍了回调,在实际的模型构建过程中,可以通过回调对程序进行调整,如可以在回调中规定,在训练准确率达到某一标准时停止并发送邮件进行通知。

7.2 函数式 API

使用 Keras 函数式 API 可以构建很多复杂的模型,如在实际中需要的多输出模型就可以用 Keras 函数式 API 构建。

7.2.1 构建简单的网络

使用函数式 API 构建的模型层可以使用 tf.keras.Model 实例来调用并返回张量。本节使用下面的例子进行介绍。

(1) 构建一个比较简单的网络模型,代码如下。

```
#以 TensorFlow 为基础构建 Keras
import tensorflow as tf
import tensorflow.keras as keras
import tensorflow.keras.layers as layers
#实例化一个 Keras 张量
inputs = tf.keras.Input(shape=(784,), name='img')
#定义网络层
h1 = layers.Dense(32, activation='relu')(inputs)
h2 = layers.Dense(32, activation='relu')(h1)
outputs = layers.Dense(10, activation='softmax')(h2)
#使用定义的张量和网络层构建模型
model = tf.keras.Model(inputs=inputs, outputs=outputs)
#输出模型详情
model.summary()
```

(2) 代码的运行结果如下。

```
Layer (type)                 Output Shape              Param #
=================================================================
img (InputLayer)             [(None, 784)]             0
_____
dense (Dense)                (None, 32)                25120
_____
dense_1 (Dense)              (None, 32)                1056
_____
dense_2 (Dense)              (None, 10)                330
```

```
=================================================================
Total params: 26,506
Trainable params: 26,506
Non-trainable params: 0
```

(3) 对模型进行训练、验证和测试，代码如下。

```python
#以 TensorFlow 为基础构建 Keras
import tensorflow as tf
import tensorflow.keras as keras
import tensorflow.keras.layers as layers
#导入数据集并定义训练集和测试集
(x_train, y_train), (x_test, y_test) = keras.datasets.mnist.load_data()
x_train = x_train.reshape(60000, 784).astype('float32') /255
x_test = x_test.reshape(10000, 784).astype('float32') /255
#实例化一个 Keras 张量
inputs = tf.keras.Input(shape=(784,), name='img')
#定义网络层
h1 = layers.Dense(32, activation='relu')(inputs)
h2 = layers.Dense(32, activation='relu')(h1)
outputs = layers.Dense(10, activation='softmax')(h2)
#构建模型
model = tf.keras.Model(inputs=inputs, outputs=outputs)
keras.utils.plot_model(model, 'mnist_model.png')
keras.utils.plot_model(model, 'model_info.png', show_shapes=True)
#初始化所有的权重参数
model.compile(optimizer=keras.optimizers.RMSprop(),
    loss='sparse_categorical_crossentropy',
    #直接使用 API
    metrics=['accuracy'])
    #训练 5 次
history = model.fit(x_train, y_train, batch_size=64, epochs=5, validation_split=0.2)
#评估模型，返回损失率和精确度
test_scores = model.evaluate(x_test, y_test, verbose=0)
print('test loss:', test_scores[0])
print('test acc:', test_scores[1])
```

(4) 代码的运行结果如下。

```
Train on 48000 samples, validate on 12000 samples
Epoch 1/5
48000/48000 [==============================] - 3s 60us/sample - loss: 0.4328 - accuracy: 0.8771 - val_loss: 0.2489 - val_accuracy: 0.9288
...
Epoch 5/5
48000/48000 [==============================] - 2s 38us/sample - loss: 0.1347 - accuracy: 0.9596 - val_loss: 0.1505 - val_accuracy: 0.9570
```

```
test loss: 0.14962136347219349
test acc: 0.9566
```

7.2.2 构建多个模型

在实际生产过程中，有时需要抽离出一个或多个图层单独构建模型，在函数 API 中，该操作十分便捷。

（1）构建模型，代码如下。

```
#以 TensorFlow 为基础构建 Keras
import tensorflow as tf
import tensorflow.keras as keras
import tensorflow.keras.layers as layers
#实例化一个 Keras 张量
encode_input = keras.Input(shape=(28,28,1), name='img')
#定义网络层 h1
h1 = layers.Conv2D(16, 3, activation='relu')(encode_input)
h1 = layers.Conv2D(32, 3, activation='relu')(h1)
h1 = layers.MaxPool2D(3)(h1)
h1 = layers.Conv2D(32, 3, activation='relu')(h1)
h1 = layers.Conv2D(16, 3, activation='relu')(h1)
encode_output = layers.GlobalMaxPool2D()(h1)
#基于网络层 h1 构建模型
encode_model = keras.Model(inputs=encode_input, outputs=encode_output, name=' encoder')
#输出模型详情
encode_model.summary()
#定义网络层 h2
h2 = layers.Reshape((4, 4, 1))(encode_output)
h2 = layers.Conv2DTranspose(16, 3, activation='relu')(h2)
h2 = layers.Conv2DTranspose(32, 3, activation='relu')(h2)
h2 = layers.UpSampling2D(3)(h2)
h2 = layers.Conv2DTranspose(16, 3, activation='relu')(h2)
decode_output = layers.Conv2DTranspose(1, 3, activation='relu')(h2)
#基于网络层 h2 构建模型
autoencoder=keras.Model(inputs=encode_input, outputs=decode_output, name='autoencoder')
#输出模型详情
autoencoder.summary()
```

（2）代码的运行结果如下。

```
Model: " encoder "
_____
Layer (type)                 Output Shape              Param #
=================================================================
img (InputLayer)             [(None, 28, 28, 1)]       0
_____
```

Layer (type)	Output Shape	Param #
conv2d (Conv2D)	(None, 26, 26, 16)	160
conv2d_1 (Conv2D)	(None, 24, 24, 32)	4640
max_pooling2d (MaxPooling2D)	(None, 8, 8, 32)	0
conv2d_2 (Conv2D)	(None, 6, 6, 32)	9248
conv2d_3 (Conv2D)	(None, 4, 4, 16)	4624
global_max_pooling2d (Global	(None, 16)	0

Total params: 18,672
Trainable params: 18,672
Non-trainable params: 0

Model: "autoencoder"

Layer (type)	Output Shape	Param #
img (InputLayer)	[(None, 28, 28, 1)]	0
conv2d (Conv2D)	(None, 26, 26, 16)	160
conv2d_1 (Conv2D)	(None, 24, 24, 32)	4640
max_pooling2d (MaxPooling2D)	(None, 8, 8, 32)	0
conv2d_2 (Conv2D)	(None, 6, 6, 32)	9248
conv2d_3 (Conv2D)	(None, 4, 4, 16)	4624
global_max_pooling2d (Global	(None, 16)	0
reshape (Reshape)	(None, 4, 4, 1)	0
conv2d_transpose (Conv2DTran	(None, 6, 6, 16)	160
conv2d_transpose_1 (Conv2DTr	(None, 8, 8, 32)	4640
up_sampling2d (UpSampling2D)	(None, 24, 24, 32)	0
conv2d_transpose_2 (Conv2DTr	(None, 26, 26, 16)	4624

```
conv2d_transpose_3 (Conv2DTr (None, 28, 28, 1)         145
=================================================================
Total params: 28,241
Trainable params: 28,241
Non-trainable params: 0
_____
```

根据第 6 章中对自编码器的理解对结果进行分析，可以发现这里通过单一图层构建了多个模型。

（3）函数 API 可以把整个模型当作一层网络来使用，代码如下。

```
#以 TensorFlow 为基础构建 Keras
import tensorflow as tf
import tensorflow.keras as keras
import tensorflow.keras.layers as layers
#实例化一个 Keras 张量
encode_input = keras.Input(shape=(28,28,1), name='img')
#定义网络层 h1
h1 = layers.Conv2D(16, 3, activation='relu')(encode_input)
h1 = layers.Conv2D(32, 3, activation='relu')(h1)
h1 = layers.MaxPool2D(3)(h1)
h1 = layers.Conv2D(32, 3, activation='relu')(h1)
h1 = layers.Conv2D(16, 3, activation='relu')(h1)
encode_output = layers.GlobalMaxPool2D()(h1)
#基于网络层 h1 构建模型
encode_model = keras.Model(inputs = encode_input, outputs=encode_output, name='encoder')
#输出模型详情
encode_model.summary()
#定义网络层 h2
h2 = layers.Reshape((4, 4, 1))(encode_output)
h2 = layers.Conv2DTranspose(16, 3, activation='relu')(h2)
h2 = layers.Conv2DTranspose(32, 3, activation='relu')(h2)
h2 = layers.UpSampling2D(3)(h2)
h2 = layers.Conv2DTranspose(16, 3, activation='relu')(h2)
decode_output = layers.Conv2DTranspose(1, 3, activation='relu')(h2)
#基于网络层 h2 构建模型
autoencoder = keras.Model(inputs=encode_input, outputs=decode_output, name='autoencoder')
#输出模型详情
autoencoder.summary()
#实例化一个 Keras 张量
autoencoder_input = keras.Input(shape=(28,28,1), name='img')
h3 = encode_model(autoencoder_input)
autoencoder_output = decode_model(h3)
#把整个模型作为一层网络使用，构建模型
autoencoder = keras.Model(inputs=autoencoder_input, outputs=autoencoder_output,
```

```
                        name='autoencoder')
autoencoder.summary()
```
（4）代码的运行结果如下。
```
Model: "encoder"
_____
Layer (type)                 Output Shape              Param #
=================================================================
src_img (InputLayer)         [(None, 28, 28, 1)]       0
_____
conv2d (Conv2D)              (None, 26, 26, 16)        160
_____
conv2d_1 (Conv2D)            (None, 24, 24, 32)        4640
_____
max_pooling2d (MaxPooling2D) (None, 8, 8, 32)          0
_____
conv2d_2 (Conv2D)            (None, 6, 6, 32)          9248
_____
conv2d_3 (Conv2D)            (None, 4, 4, 16)          4624
_____
global_max_pooling2d (Global (None, 16)                0
=================================================================
Total params: 18,672
Trainable params: 18,672
Non-trainable params: 0
_____

Model: "decoder"
_____
Layer (type)                 Output Shape              Param #
=================================================================
encoded_img (InputLayer)     [(None, 16)]              0
_____
reshape (Reshape)            (None, 4, 4, 1)           0
_____
conv2d_transpose (Conv2DTran (None, 6, 6, 16)          160
_____
conv2d_transpose_1 (Conv2DTr (None, 8, 8, 32)          4640
_____
up_sampling2d (UpSampling2D) (None, 24, 24, 32)        0
_____
conv2d_transpose_2 (Conv2DTr (None, 26, 26, 16)        4624
_____
conv2d_transpose_3 (Conv2DTr (None, 28, 28, 1)         145
=================================================================
```

```
Total params: 9,569
Trainable params: 9,569
Non-trainable params: 0
_____
Model: " autoencoder "
_____
Layer (type)                 Output Shape              Param #
=================================================================
img (InputLayer)             [(None, 28, 28, 1)]       0
_____
encoder (Model)              (None, 16)                18672
_____
decoder (Model)              (None, 28, 28, 1)         9569
=================================================================
Total params: 28,241
Trainable params: 28,241
Non-trainable params: 0
_____
```

7.2.3 两种典型的复杂网络

在实际生产过程中，仅靠简单的模型无法支撑数据识别和分析工作，本节介绍两种在实际中会使用到的复杂网络。

（1）构建一个多输入与多输出的网络，代码如下。

```
#以 TensorFlow 为基础构建 Keras
import tensorflow as tf
from tensorflow import keras
from tensorflow.keras import layers
#构建一个根据文档内容、标签和标题来预测文档优先级和执行部门的网络
num_words = 2000
num_tags = 12
num_departments = 4
#实例化 Keras 张量，并将其作为输入
body_input = keras.Input(shape=(None,), name='body')
title_input = keras.Input(shape=(None,), name='title')
tag_input = keras.Input(shape=(num_tags,), name='tag')
#定义嵌入层
body_feat = layers.Embedding(num_words, 64)(body_input)
title_feat = layers.Embedding(num_words, 64)(title_input)
#定义特征提取层
body_feat = layers.LSTM(32)(body_feat)
title_feat = layers.LSTM(128)(title_feat)
features = layers.concatenate([title_feat,body_feat, tag_input])
```

```
#定义分类层
priority_pred = layers.Dense(1, activation='sigmoid', name='priority')(features)
department_pred = layers.Dense(num_departments, activation='softmax', name=' department')
(features)
#构建模型
model = keras.Model(inputs=[body_input, title_input, tag_input],
    outputs=[priority_pred, department_pred])
#输出模型详情
model.summary()
#初始化所有的权重参数
model.compile(optimizer=keras.optimizers.RMSprop(1e-3),
    loss={'priority': 'binary_crossentropy',
          'department': 'categorical_crossentropy'},
    loss_weights=[1., 0.2])
#导入NumPy模块
import numpy as np
#载入输入数据并区分数据
title_data = np.random.randint(num_words, size=(1280, 10))
body_data = np.random.randint(num_words, size=(1280, 100))
tag_data = np.random.randint(2, size=(1280, num_tags)).astype('float32')
#对数据标签进行处理
priority_label = np.random.random(size=(1280, 1))
department_label = np.random.randint(2, size=(1280, num_departments))
#制订训练计划
history = model.fit(
    {'title': title_data, 'body':body_data, 'tag':tag_data},
    {'priority':priority_label, 'department':department_label},
    batch_size=32,
    epochs=5
)
```

(2)代码的运行结果如下。

```
Model: "model"
_____
Layer (type)              Output Shape          Param #      Connected to
==========================================================================
title (InputLayer)        [(None, None)]        0
_____
body (InputLayer)         [(None, None)]        0
_____
embedding_1 (Embedding)   (None, None, 64)      128000       title[0][0]
_____
embedding (Embedding)     (None, None, 64)      128000       body[0][0]
_____
```

```
lstm_1 (LSTM)              (None, 128)      98816    embedding_1[0][0]
_____
lstm (LSTM)                (None, 32)       12416    embedding[0][0]
_____
tag (InputLayer)           [(None, 12)]     0
_____
concatenate (Concatenate)  (None, 172)      0        lstm_1[0][0]
                                                     lstm[0][0]
                                                     tag[0][0]
_____
priority (Dense)           (None, 1)        173      concatenate[0][0]
_____
department (Dense)         (None, 4)        692      concatenate[0][0]
==================================================================
Total params: 368,097
Trainable params: 368,097
Non-trainable params: 0
_____
Train on 1280 samples
Epoch 1/5
1280/1280 [==============================] - 7s 5ms/sample - loss: 1.3151 - priority_loss: 0.7020 - department_loss: 3.0657
...
Epoch 5/5
1280/1280 [==============================] - 3s 2ms/sample - loss: 1.2656 - priority_loss: 0.6971 - department_loss: 2.8423
```

（3）构建一个小型残差网络，残差网络的特点是容易优化，且能够通过增加深度来提高准确率。其内部的残差块使用跳跃连接，解决了在深度神经网络中增加深度带来的梯度消失问题。其代码如下。

```
#以 TensorFlow 为基础构建 Keras
import ssl
import tensorflow as tf
import matplotlib.pyplot as plt
from tensorflow import keras
from tensorflow.keras import layers
#载入能够访问 HTTPS 的网络证书
ssl._create_default_https_context = ssl._create_unverified_context
#实例化 Keras 张量，并将其作为输入
inputs = keras.Input(shape=(32,32,3), name='img')
#定义网络层 h1
h1 = layers.Conv2D(32, 3, activation='relu')(inputs)
h1 = layers.Conv2D(64, 3, activation='relu')(h1)
block1_out = layers.MaxPooling2D(3)(h1)
```

```
#定义网络层h2
h2 = layers.Conv2D(64, 3, activation='relu', padding='same')(block1_out)
h2 = layers.Conv2D(64, 3, activation='relu', padding='same')(h2)
block2_out = layers.add([h2, block1_out])
#定义网络层h3
h3 = layers.Conv2D(64, 3, activation='relu', padding='same')(block2_out)
h3 = layers.Conv2D(64, 3, activation='relu', padding='same')(h3)
block3_out = layers.add([h3, block2_out])
#定义网络层h4
h4 = layers.Conv2D(64, 3, activation='relu')(block3_out)
h4 = layers.GlobalMaxPool2D()(h4)
h4 = layers.Dense(256, activation='relu')(h4)
h4 = layers.Dropout(0.5)(h4)
outputs = layers.Dense(10, activation='softmax')(h4)
#定义模型
model = keras.Model(inputs, outputs)
#输出模型详情
model.summary()
#导入数据集并划分训练集和测试集
(x_train, y_train), (x_test, y_test) = keras.datasets.cifar10.load_data()
x_train = x_train.astype('float32') / 255
x_test = y_train.astype('float32') / 255
y_train = keras.utils.to_categorical(y_train, 10)
y_test = keras.utils.to_categorical(y_test, 10)
#初始化所有的权重参数
model.compile(optimizer=keras.optimizers.RMSprop(1e-3),
    loss='categorical_crossentropy', metrics=['acc'])
#制订训练计划并进行训练
model.fit(x_train, y_train, batch_size=64, epochs=1, validation_split=0.2)
```

（4）代码的运行结果如下。

```
Model: " model "
_____
Layer (type)                  Output Shape        Param #     Connected to
==========================================================================
img (InputLayer)              [(None, 32, 32, 3)]  0

conv2d (Conv2D)               (None, 30, 30, 32)   896         img[0][0]

conv2d_1 (Conv2D)             (None, 28, 28, 64)   18496       conv2d[0][0]

max_pooling2d (MaxPooling2D)  (None, 9, 9, 64)     0           conv2d_1[0][0]

conv2d_2 (Conv2D) (None, 9, 9, 64)    36928       max_pooling2d[0][0]
```

```
conv2d_3 (Conv2D)               (None, 9, 9, 64)     36928       conv2d_2[0][0]
_____
add (Add)                       (None, 9, 9, 64)     0           conv2d_3[0][0]
                                                                 max_pooling2d[0][0]
_____
conv2d_4 (Conv2D)               (None, 9, 9, 64)     36928       add[0][0]
_____
conv2d_5 (Conv2D)               (None, 9, 9, 64)     36928       conv2d_4[0][0]
_____
add_1 (Add)                     (None, 9, 9, 64)     0           conv2d_5[0][0]
                                                                 add[0][0]
_____
conv2d_6 (Conv2D)               (None, 7, 7, 64)     36928       add_1[0][0]
_____
global_max_pooling2d (GlobalMax (None, 64)           0           conv2d_6[0][0]
_____
dense (Dense)        (None, 256)                     16640       global_max_pooling2d[0][0]
_____
dropout (Dropout)               (None, 256)          0           dense[0][0]
_____
dense_1 (Dense)                 (None, 10)           2570        dropout[0][0]
================================================================================
Total params: 223,242
Trainable params: 223,242
Non-trainable params: 0
_____
Train on 40000 samples, validate on 10000 samples
40000/40000 [==============================] - 361s 9ms/sample - loss: 1.8662 - acc: 0.2974 - val_loss: 1.4558 - val_acc: 0.4581
```

7.3 使用 Keras 自定义网络层和模型

Keras 提供了丰富的 API，可以协助构建多样的网络和模型，本节介绍如何使用 Keras 自定义网络层和模型。

7.3.1 构建简单网络

本节根据简单网络的构建及参数的调整介绍如何使用 Keras 构建网络，使读者对使用 Keras 构建网络有初步的认识。

（1）构建一个简单的网络，代码如下。

```
#引入 TensorFlow 模块并基于 TensorFlow 构建 Keras
```

```
from __future__ import absolute_import, division, print_function
import tensorflow as tf
import tensorflow.keras as keras
import tensorflow.keras.layers as layers
#定义网络层,实际上就是设置网络权重和从输出到输入的计算过程
class MyLayer(layers.Layer):
    def __init__(self, input_dim=32, unit=32):
        super(MyLayer, self).__init__()
        w_init = tf.random_normal_initializer()
        self.weight = tf.Variable(initial_value=w_init(
            shape=(input_dim, unit), dtype=tf.float32), trainable=True)
        b_init = tf.zeros_initializer()
        self.bias = tf.Variable(initial_value=b_init(
            shape=(unit,), dtype=tf.float32), trainable=True)
    def call(self, inputs):
        return tf.matmul(inputs, self.weight) + self.bias
#创建一个将所有元素都设置为1的张量
x = tf.ones((3, 5))
my_layer = MyLayer(5, 4)
out = my_layer(x)
print(out)
```

(2)代码的运行结果如下。

```
tf.Tensor(
[[-0.06724034  0.08593177  0.05762753 -0.08204076]
 [-0.06724034  0.08593177  0.05762753 -0.08204076]
 [-0.06724034  0.08593177  0.05762753 -0.08204076]], shape=(3, 4), dtype=float32)
```

(3)可以发现,图层会自动跟踪权重 w 和 b,下面使用 add_weight 构建权重网络,代码如下。

```
#引入 TensorFlow 模块并基于 TensorFlow 构建 Keras
from __future__ import absolute_import, division, print_function
import tensorflow as tf
import tensorflow.keras as keras
import tensorflow.keras.layers as layers
#销毁当前的 TF 图并创建一个新图,该操作有助于避免旧模型或图层混乱
tf.keras.backend.clear_session()
#使用 add_weight 构建权重网络
class MyLayer(layers.Layer):
    def __init__(self, input_dim=32, unit=32):
        super(MyLayer, self).__init__()
        self.weight = self.add_weight(shape=(input_dim, unit),
            initializer=keras.initializers.RandomNormal(), trainable=True)
        self.bias = self.add_weight(shape=(unit,),
            initializer=keras.initializers.Zeros(), trainable=True)
```

```
        def call(self, inputs):
            return tf.matmul(inputs, self.weight) + self.bias
    #创建一个将所有元素都设置为1的张量
    x = tf.ones((3, 5))
    my_layer = MyLayer(5, 4)
    out = my_layer(x)
    print(out)
```

（4）代码的运行结果如下。

```
    tf.Tensor(
    [[ 0.21920022 -0.06261671 -0.03179146 -0.0403799 ]
     [ 0.21920022 -0.06261671 -0.03179146 -0.0403799 ]
     [ 0.21920022 -0.06261671 -0.03179146 -0.0403799 ]], shape=(3, 4), dtype=float32)
```

（5）在训练过程中，也可以将网络设置为不可训练的权重网络，代码如下。

```
    #引入 TensorFlow 模块并基于 TensorFlow 构建 Keras
    from __future__ import absolute_import, division, print_function
    import tensorflow as tf
    import tensorflow.keras as keras
    import tensorflow.keras.layers as layers
    #销毁当前的 TF 图并创建一个新图，该操作有助于避免旧模或图层混乱
    tf.keras.backend.clear_session()
    #使用 add_weight 构建权重网络
    class MyLayer(layers.Layer):
        def __init__(self, input_dim=32, unit=32):
            super(MyLayer, self).__init__()
            self.weight = self.add_weight(shape=(input_dim, unit),
                initializer=keras.initializers.RandomNormal(), trainable=True)
            self.bias = self.add_weight(shape=(unit,),
                initializer=keras.initializers.Zeros(), trainable=True)
        def call(self, inputs):
            return tf.matmul(inputs, self.weight) + self.bias
    class AddLayer(layers.Layer):
        def __init__(self, input_dim=32):
            super(AddLayer, self).__init__()
            self.sum = self.add_weight(shape=(input_dim,),
                initializer=keras.initializers.Zeros(), trainable=False)
        def call(self, inputs):
            self.sum.assign_add(tf.reduce_sum(inputs, axis=0))
            return self.sum
    #创建一个将所有元素都设置为1的张量
    x = tf.ones((3, 3))
    my_layer = AddLayer(3)
    out = my_layer(x)
    print(out.numpy())
```

```
out = my_layer(x)
print(out.numpy())
print('weight:', my_layer.weights)
print('non-trainable weight:', my_layer.non_trainable_weights)
print('trainable weight:', my_layer.trainable_weights)
```

（6）代码的运行结果如下。

```
[3. 3. 3.]
[6. 6. 6.]
weight: [<tf.Variable 'Variable:0' shape=(3,) dtype=float32, numpy=array([6., 6., 6.], dtype=float32)>]
non-trainable weight: [<tf.Variable 'Variable:0' shape=(3,) dtype=float32, numpy=array([6., 6., 6.], dtype=float32)>]
trainable weight: []
```

（7）在实际的机器学习过程中，有时会出现不知道网络维度的情况，下面对这种情况进行演示。通过重写 build() 函数，用已获得的 shape 构建相应的网络，代码如下。

```
#引入 TensorFlow 模块并基于 TensorFlow 构建 Keras
from __future__ import absolute_import, division, print_function
import tensorflow as tf
import tensorflow.keras as keras
import tensorflow.keras.layers as layers
#销毁当前的 TF 图并创建一个新图，该操作有助于避免旧模型或图层混乱
tf.keras.backend.clear_session()
#重写 build() 函数，用获得的 shape 构建相应的网络
class MyLayer(layers.Layer):
    def __init__(self, unit=32):
        super(MyLayer, self).__init__()
        self.unit = unit
    def build(self, input_shape):
        self.weight = self.add_weight(shape=(input_shape[-1], self.unit),
            initializer=keras.initializers.RandomNormal(), trainable=True)
        self.bias = self.add_weight(shape=(self.unit,),
            initializer=keras.initializers.Zeros(), trainable=True)
    def call(self, inputs):
        return tf.matmul(inputs, self.weight) + self.bias
#使用自定义方法构建网络
my_layer = MyLayer(3)
x = tf.ones((3, 5))
out = my_layer(x)
print(out)
my_layer = MyLayer(3)
#创建一个将所有元素都设置为 1 的张量
x = tf.ones((2, 2))
#使用重新定义的网络
```

```
out = my_layer(x)
print(out)
```
(8) 代码的运行结果如下。
```
tf.Tensor(
[[0.01456478 0.07924593 0.02756532]
 [0.01456478 0.07924593 0.02756532]
 [0.01456478 0.07924593 0.02756532]], shape=(3, 3), dtype=float32)
tf.Tensor(
[[ 0.01749954 -0.04188586  0.1059282 ]
 [ 0.01749954 -0.04188586  0.1059282 ]], shape=(2, 3), dtype=float32)
```

7.3.2 构建自定义模型

在实际生产过程中，通常使用 Layer 类定义内部计算块，使用 Model 类定义外部模型，即要训练的对象。本节介绍如何构建一个自定义模型。

(1) 先构建一个变分自编码器（VAE），代码如下。

```
#引入 TensorFlow 模块并基于 TensorFlow 构建 Keras
from __future__ import absolute_import, division, print_function
import tensorflow as tf
import tensorflow.keras as keras
import tensorflow.keras.layers as layers
#销毁当前的 TF 图并创建一个新图，该操作有助于避免旧模型或图层混乱
tf.keras.backend.clear_session()
#采样网络
class Sampling(layers.Layer):
    def call(self, inputs):
        z_mean, z_log_var = inputs
        batch = tf.shape(z_mean)[0]
        dim = tf.shape(z_mean)[1]
        epsilon = tf.keras.backend.random_normal(shape=(batch, dim))
        return z_mean + tf.exp(0.5 * z_log_var) * epsilon
#编码器
class Encoder(layers.Layer):
    def __init__(self, latent_dim=32,
                 intermediate_dim=64, name='encoder', **kwargs):
        super(Encoder, self).__init__(name=name, **kwargs)
        self.dense_proj = layers.Dense(intermediate_dim, activation='relu')
        self.dense_mean = layers.Dense(latent_dim)
        self.dense_log_var = layers.Dense(latent_dim)
        self.sampling = Sampling()
    def call(self, inputs):
        h1 = self.dense_proj(inputs)
        z_mean = self.dense_mean(h1)
```

```python
        z_log_var = self.dense_log_var(h1)
        z = self.sampling((z_mean, z_log_var))
        return z_mean, z_log_var, z
#解码器
class Decoder(layers.Layer):
    def __init__(self, original_dim, intermediate_dim=64, name='decoder', **kwargs):
        super(Decoder, self).__init__(name=name, **kwargs)
        self.dense_proj = layers.Dense(intermediate_dim, activation='relu')
        self.dense_output = layers.Dense(original_dim, activation='sigmoid')
    def call(self, inputs):
        h1 = self.dense_proj(inputs)
        return self.dense_output(h1)
#变分自编码器
class VAE(tf.keras.Model):
    def __init__(self, original_dim, latent_dim=32,
            intermediate_dim=64, name='encoder', **kwargs):
        super(VAE, self).__init__(name=name, **kwargs)
        self.original_dim = original_dim
        self.encoder = Encoder(latent_dim=latent_dim,
            intermediate_dim=intermediate_dim)
        self.decoder = Decoder(original_dim=original_dim,
            intermediate_dim=intermediate_dim)
    def call(self, inputs):
        z_mean, z_log_var, z = self.encoder(inputs)
        reconstructed = self.decoder(z)
        kl_loss = -0.5 * tf.reduce_sum(
            z_log_var - tf.square(z_mean) - tf.exp(z_log_var) + 1)
        self.add_loss(kl_loss)
        return reconstructed
#载入数据并划分训练集和测试集
(x_train, _), _ = tf.keras.datasets.mnist.load_data()
x_train = x_train.reshape(60000, 784).astype('float32') / 255
#使用自定义变分自编码器并初始化权重等参数
vae = VAE(784, 32, 64)
optimizer = tf.keras.optimizers.Adam(learning_rate=1e-3)
vae.compile(optimizer, loss=tf.keras.losses.MeanSquaredError())
#训练模型
vae.fit(x_train, x_train, epochs=3, batch_size=64)
```

(2) 代码的运行结果如下。

```
Train on 60000 samples
Epoch 1/3
60000/60000 [==============================] - 5s 75us/sample - loss: 0.8294
Epoch 2/3
```

```
60000/60000 [==============================] - 5s 79us/sample - loss: 0.0692
Epoch 3/3
60000/60000 [==============================] - 4s 62us/sample - loss: 0.0679
```

（3）重新编写训练方法，并替换默认的训练方法，代码如下。

```
#引入TensorFlow模块并基于TensorFlow构建Keras
from __future__ import absolute_import, division, print_function
import tensorflow as tf
import tensorflow.keras as keras
import tensorflow.keras.layers as layers
#销毁当前的TF图并创建一个新图，该操作有助于避免旧模型或图层混乱
tf.keras.backend.clear_session()
#采样网络
class Sampling(layers.Layer):
    def call(self, inputs):
        z_mean, z_log_var = inputs
        batch = tf.shape(z_mean)[0]
        dim = tf.shape(z_mean)[1]
        epsilon = tf.keras.backend.random_normal(shape=(batch, dim))
        return z_mean + tf.exp(0.5 * z_log_var) * epsilon
#编码器
class Encoder(layers.Layer):
    def __init__(self, latent_dim=32,
        intermediate_dim=64, name='encoder', **kwargs):
        super(Encoder, self).__init__(name=name, **kwargs)
        self.dense_proj = layers.Dense(intermediate_dim, activation='relu')
        self.dense_mean = layers.Dense(latent_dim)
        self.dense_log_var = layers.Dense(latent_dim)
        self.sampling = Sampling()
    def call(self, inputs):
        h1 = self.dense_proj(inputs)
        z_mean = self.dense_mean(h1)
        z_log_var = self.dense_log_var(h1)
        z = self.sampling((z_mean, z_log_var))
        return z_mean, z_log_var, z
#解码器
class Decoder(layers.Layer):
    def __init__(self, original_dim,
        intermediate_dim=64, name='decoder', **kwargs):
        super(Decoder, self).__init__(name=name, **kwargs)
        self.dense_proj = layers.Dense(intermediate_dim, activation='relu')
        self.dense_output = layers.Dense(original_dim, activation='sigmoid')
    def call(self, inputs):
```

```python
        h1 = self.dense_proj(inputs)
        return self.dense_output(h1)
#变分自编码器
class VAE(tf.keras.Model):
    def __init__(self, original_dim, latent_dim=32,
        intermediate_dim=64, name='encoder', **kwargs):
        super(VAE, self).__init__(name=name, **kwargs)

        self.original_dim = original_dim
        self.encoder = Encoder(latent_dim=latent_dim,
            intermediate_dim=intermediate_dim)
        self.decoder = Decoder(original_dim=original_dim,
            intermediate_dim=intermediate_dim)
    def call(self, inputs):
        z_mean, z_log_var, z = self.encoder(inputs)
        reconstructed = self.decoder(z)
        kl_loss = -0.5 * tf.reduce_sum(
            z_log_var - tf.square(z_mean) - tf.exp(z_log_var) + 1)
        self.add_loss(kl_loss)
        return reconstructed
#导入数据集并划分训练集和测试集
(x_train, _), _ = tf.keras.datasets.mnist.load_data()
x_train = x_train.reshape(60000, 784).astype('float32') / 255
#使用变分自编码器构建模型并初始化学习率参数
vae = VAE(784,32,64)
optimizer = tf.keras.optimizers.Adam(learning_rate=1e-3)
train_dataset = tf.data.Dataset.from_tensor_slices(x_train)
train_dataset = train_dataset.shuffle(buffer_size=1024).batch(64)
original_dim = 784
vae = VAE(original_dim, 64, 32)
optimizer = tf.keras.optimizers.Adam(learning_rate=1e-3)
mse_loss_fn = tf.keras.losses.MeanSquaredError()
loss_metric = tf.keras.metrics.Mean()
#迭代
for epoch in range(3):
    print('Start of epoch %d' % (epoch,))
#取出每个batch的数据并训练
for step, x_batch_train in enumerate(train_dataset):
    with tf.GradientTape() as tape:
        reconstructed = vae(x_batch_train)
        #计算reconstruction loss
        loss = mse_loss_fn(x_batch_train, reconstructed)
        loss += sum(vae.losses)
        #添加KLD regularization loss
```

```
        grads = tape.gradient(loss, vae.trainable_variables)
        optimizer.apply_gradients(zip(grads, vae.trainable_variables))
        loss_metric(loss)
        if step % 100 == 0:
            print('step %s: mean loss = %s' % (step, loss_metric.result()))
```
(4)代码的运行结果如下。
```
Start of epoch 0
Start of epoch 1
Start of epoch 2
step 0: mean loss = tf.Tensor(263.8553, shape=(), dtype=float32)
...
step 900: mean loss = tf.Tensor(1.0207621, shape=(), dtype=float32)
```

7.4 Keras 训练模型

在构建完网络和模型后,需要对加载数据的模型进行训练,在训练的过程中需要不断调整各种参数,本节对 Keras 训练模型进行介绍。

7.4.1 常见模型的训练流程

本节对常见模型的训练流程进行介绍。
(1)构建和训练模型的代码如下。
```
#引入 TensorFlow 模块并基于 TensorFlow 构建 Keras
from __future__ import absolute_import, division, print_function
import tensorflow as tf
import tensorflow.keras as keras
import tensorflow.keras.layers as layers
#销毁当前的 TF 图并创建一个新图,该操作有助于避免旧模型或图层混乱
tf.keras.backend.clear_session()
#构建模型
inputs = keras.Input(shape=(784,), name='mnist_input')
h1 = layers.Dense(64, activation='relu')(inputs)
h1 = layers.Dense(64, activation='relu')(h1)
outputs = layers.Dense(10, activation='softmax')(h1)
model = keras.Model(inputs, outputs)
#初始化权重等参数
model.compile(optimizer=keras.optimizers.RMSprop(),
    loss=keras.losses.SparseCategoricalCrossentropy(),
    metrics=[keras.metrics.SparseCategoricalAccuracy()])
#导入数据集并划分训练集和测试集
(x_train, y_train), (x_test, y_test) = keras.datasets.mnist.load_data()
```

```
x_train = x_train.reshape(60000, 784).astype('float32') /255
x_test = x_test.reshape(10000, 784).astype('float32') /255
x_val = x_train[-10000:]
y_val = y_train[-10000:]
x_train = x_train[:-10000]
y_train = y_train[:-10000]
#训练模型
history = model.fit(x_train, y_train, batch_size=64, epochs=3,
    validation_data=(x_val, y_val))
print('history:')
print(history.history)
#对模型进行评估
result = model.evaluate(x_test, y_test, batch_size=128)
print('evaluate:')
print(result)
pred = model.predict(x_test[:2])
print('predict:')
print(pred)
```

（2）代码的运行结果如下。

```
Train on 50000 samples, validate on 10000 samples
Epoch 1/3
50000/50000 [==============================] - 3s 62us/sample - loss: 0.3406 - sparse_categorical_accuracy: 0.9022 - val_loss: 0.2137 - val_sparse_categorical_accuracy: 0.9362
...
Epoch 3/3
50000/50000 [==============================] - 2s 44us/sample - loss: 0.1187 - sparse_categorical_accuracy: 0.9646 - val_loss: 0.1183 - val_sparse_categorical_accuracy: 0.9672
history:
    {'loss':    [0.34056663007736204,    0.161259304510355,    0.11870563844621182],
'sparse_categorical_accuracy': [0.90224, 0.95194, 0.96462], 'val_loss': [0.21368005533218384,
0.134373027305305, 0.11827467557229102], 'val_sparse_categorical_accuracy': [0.9362, 0.9625,
0.9672]}
    10000/1 [==============================================================] - 0s 13us/sample - loss: 0.0608 - sparse_categorical_accuracy: 0.9637
evaluate:
[0.11747951722219586, 0.9637]
predict:
[[3.5755113e-07 3.7563577e-08 4.3881464e-05 5.6507859e-05 4.8591668e-09
  1.0834727e-06 4.3809325e-13 9.9987602e-01 1.9619726e-07 2.1969072e-05]
 [9.4039142e-06 5.9000308e-06 9.9991047e-01 6.9690250e-05 6.8645395e-11
  1.5047639e-06 1.3832307e-06 1.2749730e-09 1.5155335e-06 7.5686174e-10]]
```

7.4.2 自定义指标

可以通过自定义损失率等指标对基础模型进行优化,本节对自定义损失率等指标的方法进行介绍。

(1) 自定义指标只需继承 metric 类,并重写初始化函数_init_(self),本例采用定义网络层的方式添加网络 loss,示例代码如下。

```
#引入 TensorFlow 模块并基于 TensorFlow 构建 Keras
from __future__ import absolute_import, division, print_function
import tensorflow as tf
import tensorflow.keras as keras
import tensorflow.keras.layers as layers
#销毁当前的 TF 图并创建一个新图,该操作有助于避免旧模型或图层混乱
tf.keras.backend.clear_session()
#导入数据集并划分训练集和测试集
(x_train, y_train), (x_test, y_test) = keras.datasets.mnist.load_data()
x_train = x_train.reshape(60000, 784).astype('float32') /255
x_test = x_test.reshape(10000, 784).astype('float32') /255
x_val = x_train[-10000:]
y_val = y_train[-10000:]
x_train = x_train[:-10000]
y_train = y_train[:-10000]
#采用定义网络层的方式添加网络 loss
class ActivityRegularizationLayer(layers.Layer):
    def call(self, inputs):
        self.add_loss(tf.reduce_sum(inputs) * 0.1)
        return inputs
#定义 Keras 张量,并将其作为输入
inputs = keras.Input(shape=(784,), name='mnist_input')
#定义网络层 h1
h1 = layers.Dense(64, activation='relu')(inputs)
h1 = ActivityRegularizationLayer()(h1)
h1 = layers.Dense(64, activation='relu')(h1)
outputs = layers.Dense(10, activation='softmax')(h1)
#构建模型
model = keras.Model(inputs, outputs)
#初始化权重、损失率等参数
model.compile(optimizer=keras.optimizers.RMSprop(),
    loss=keras.losses.SparseCategoricalCrossentropy(),
    metrics=[keras.metrics.SparseCategoricalAccuracy()])
#训练模型
model.fit(x_train, y_train, batch_size=32, epochs=1)
```

（2）代码的运行结果如下。

```
Train on 50000 samples
50000/50000 [==============================] - 5s 94us/sample - loss: 2.3489 - sparse_categorical_accuracy: 0.1134
```

（3）采用定义网络层的方式添加要统计的 metric，示例代码如下。

```
#引入TensorFlow模块并基于TensorFlow构建Keras
from __future__ import absolute_import, division, print_function
import tensorflow as tf
import tensorflow.keras as keras
import tensorflow.keras.layers as layers
#销毁当前的TF图并创建一个新图，该操作有助于避免旧模型或图层混乱
tf.keras.backend.clear_session()
#导入数据集并划分训练集和测试集
(x_train, y_train), (x_test, y_test) = keras.datasets.mnist.load_data()
x_train = x_train.reshape(60000, 784).astype('float32') /255
x_test = x_test.reshape(10000, 784).astype('float32') /255
x_val = x_train[-10000:]
y_val = y_train[-10000:]
x_train = x_train[:-10000]
y_train = y_train[:-10000]
#采用定义网络层的方式添加要统计的metric
class MetricLoggingLayer(layers.Layer):
    def call(self, inputs):
        self.add_metric(keras.backend.std(inputs),
            name='std_of_activation', aggregation='mean')
        return inputs
#定义Keras张量，并将其作为输入
inputs = keras.Input(shape=(784,), name='mnist_input')
#定义网络层h1
h1 = layers.Dense(64, activation='relu')(inputs)
h1 = MetricLoggingLayer()(h1)
h1 = layers.Dense(64, activation='relu')(h1)
outputs = layers.Dense(10, activation='softmax')(h1)
#构建模型
model = keras.Model(inputs, outputs)
#初始化权重、损失率等参数
model.compile(optimizer=keras.optimizers.RMSprop(),
    loss=keras.losses.SparseCategoricalCrossentropy(),
    metrics=[keras.metrics.SparseCategoricalAccuracy()])
#训练模型
model.fit(x_train, y_train, batch_size=32, epochs=1)
```

（4）代码的运行结果如下。

```
Train on 50000 samples
```

```
50000/50000 [==============================] - 5s 99us/sample - loss: 0.3021 - sparse_categorical_accuracy: 0.9123 - std_of_activation: 1.0142
```

（5）TensorFlow 对参数的调整十分灵活。可以在模型上进行参数调整，也可以通过定义网络层的方式添加要统计的 metric，代码如下。

```
#引入 TensorFlow 模块并基于 TensorFlow 构建 Keras
from __future__ import absolute_import, division, print_function
import tensorflow as tf
import tensorflow.keras as keras
import tensorflow.keras.layers as layers
#销毁当前的 TF 图并创建一个新图，该操作有助于避免旧模型或图层混乱
tf.keras.backend.clear_session()
#导入数据集并划分训练集和测试集
(x_train, y_train), (x_test, y_test) = keras.datasets.mnist.load_data()
x_train = x_train.reshape(60000, 784).astype('float32') /255
x_test = x_test.reshape(10000, 784).astype('float32') /255
x_val = x_train[-10000:]
y_val = y_train[-10000:]
x_train = x_train[:-10000]
y_train = y_train[:-10000]
#定义一个网络层的类
class MetricLoggingLayer(layers.Layer):
    def call(self, inputs):
        self.add_metric(keras.backend.std(inputs),
            name='std_of_activation', aggregation='mean')
        return inputs
#定义 Keras 张量，并将其作为输入
inputs = keras.Input(shape=(784,), name='mnist_input')
#定义网络层 h1 和 h2
h1 = layers.Dense(64, activation='relu')(inputs)
h2 = layers.Dense(64, activation='relu')(h1)
outputs = layers.Dense(10, activation='softmax')(h2)
#构建模型
model = keras.Model(inputs, outputs)
#在已经构建的模型中添加 metric
model.add_metric(keras.backend.std(inputs),
    name='std_of_activation', aggregation='mean')
#在已经构建的模型中添加 loss
model.add_loss(tf.reduce_sum(h1)*0.1)
#初始化权重、损失率等参数
model.compile(optimizer=keras.optimizers.RMSprop(),
    loss=keras.losses.SparseCategoricalCrossentropy(),
    metrics=[keras.metrics.SparseCategoricalAccuracy()])
```

```
#训练模型
model.fit(x_train, y_train, batch_size=32, epochs=1)
```
（6）代码的运行结果如下。
```
Train on 50000 samples
50000/50000 [==============================] - 6s 119us/sample - loss: 2.3579 - sparse_categorical_accuracy: 0.1125 - std_of_activation: 0.3083
```
（7）对于某些大型数据集来说，需要划分验证数据。在这种情况下，可以使用 validation_split，令 validation_split 为 0.3，示例代码如下。
```
#引入 TensorFlow 模块并基于 TensorFlow 构建 Keras
from __future__ import absolute_import, division, print_function
import tensorflow as tf
import tensorflow.keras as keras
import tensorflow.keras.layers as layers
#销毁当前的 TF 图并创建一个新图，该操作有助于避免旧模型或图层混乱
tf.keras.backend.clear_session()
#导入数据集并划分训练集和测试集
(x_train, y_train), (x_test, y_test) = keras.datasets.mnist.load_data()
x_train = x_train.reshape(60000, 784).astype('float32') /255
x_test = x_test.reshape(10000, 784).astype('float32') /255
x_val = x_train[-10000:]
y_val = y_train[-10000:]
x_train = x_train[:-10000]
y_train = y_train[:-10000]
#定义一个网络层的类
class MetricLoggingLayer(layers.Layer):
    def call(self, inputs):
        self.add_metric(keras.backend.std(inputs),
            name='std_of_activation', aggregation='mean')
        return inputs
#定义 Keras 张量，并将其作为输入
inputs = keras.Input(shape=(784,), name='mnist_input')
#定义网络层 h1 和 h2
h1 = layers.Dense(64, activation='relu')(inputs)
h2 = layers.Dense(64, activation='relu')(h1)
outputs = layers.Dense(10, activation='softmax')(h2)
#构建模型
model = keras.Model(inputs, outputs)
#在已经构建的模型中添加 metric
model.add_metric(keras.backend.std(inputs),
    name='std_of_activation', aggregation='mean')
#在已经构建的模型中添加 loss
model.add_loss(tf.reduce_sum(h1)*0.1)
```

```
#初始化权重、损失率等参数
model.compile(optimizer=keras.optimizers.RMSprop(),
    loss=keras.losses.SparseCategoricalCrossentropy(),
    metrics=[keras.metrics.SparseCategoricalAccuracy()])
#训练模型
model.fit(x_train, y_train, batch_size=32, epochs=1, validation_split=0.3)
```

说明：validation_split 用于不提供验证集的情况，能够按一定比例从训练集中取出一部分并将其作为验证集，本例中该比例为30%。

（8）代码的运行结果如下。

```
Train on 35000 samples, validate on 15000 samples
35000/35000 [==============================] - 5s 157us/sample - loss: 2.3698 - sparse_categorical_accuracy: 0.1145 - std_of_activation: 0.3085 - val_loss: 2.3020 - val_sparse_categorical_accuracy: 0.1117 - val_std_of_activation: 0.3073
```

7.4.3 自定义训练和验证循环

可以通过自定义训练和验证循环对基础模型进行优化。

（1）构造训练和循环，代码如下。

```
#引入TensorFlow模块并基于TensorFlow构建Keras
from __future__ import absolute_import, division, print_function
import tensorflow as tf
import tensorflow.keras as keras
import tensorflow.keras.layers as layers
#销毁当前的TF图并创建一个新图，该操作有助于避免旧模型或图层混乱
tf.keras.backend.clear_session()
#导入数据集并划分训练集和测试集
(x_train, y_train), (x_test, y_test) = keras.datasets.mnist.load_data()
x_train = x_train.reshape(60000, 784).astype('float32') /255
x_test = x_test.reshape(10000, 784).astype('float32') /255
x_val = x_train[-10000:]
y_val = y_train[-10000:]
x_train = x_train[:-10000]
y_train = y_train[:-10000]
#构建一个全连接网络
inputs = keras.Input(shape=(784,), name='digits')
x = layers.Dense(64, activation='relu', name='dense_1')(inputs)
x = layers.Dense(64, activation='relu', name='dense_2')(x)
outputs = layers.Dense(10, activation='softmax', name='predictions')(x)
model = keras.Model(inputs=inputs, outputs=outputs)
#建立优化器
optimizer = keras.optimizers.SGD(learning_rate=1e-3)
#损失函数
```

```python
loss_fn = keras.losses.SparseCategoricalCrossentropy()
#准备训练数据
batch_size = 64
train_dataset = tf.data.Dataset.from_tensor_slices((x_train, y_train))
train_dataset = train_dataset.shuffle(buffer_size=1024).batch(batch_size)
#构造循环
for epoch in range(3):
    print('epoch: ', epoch)
    for step, (x_batch_train, y_batch_train) in enumerate(train_dataset):
        #计算梯度
        with tf.GradientTape() as tape:
            logits = model(x_batch_train)
            loss_value = loss_fn(y_batch_train, logits)
        grads = tape.gradient(loss_value, model.trainable_variables)
        optimizer.apply_gradients(zip(grads, model.trainable_variables))
        if step % 200 == 0:
            print('Training loss (for one batch) at step %s: %s'
                % (step, float(loss_value)))
            print('Seen so far: %s samples' % ((step + 1) * 64))
```

(2) 代码的运行结果如下。

```
epoch:  0
Training loss (for one batch) at step 0: 2.3696329593658447
Seen so far: 64 samples
Training loss (for one batch) at step 200: 2.279789924621582
Seen so far: 12864 samples
Training loss (for one batch) at step 400: 2.1937031745910645
Seen so far: 25664 samples
Training loss (for one batch) at step 600: 2.1733367443084717
Seen so far: 38464 samples
epoch:  1
Training loss (for one batch) at step 0: 2.1400647163391113
Seen so far: 64 samples
...
Training loss (for one batch) at step 600: 1.8723350763320923
Seen so far: 38464 samples
epoch:  2
Training loss (for one batch) at step 0: 1.8295190334320068
Seen so far: 64 samples
...
Training loss (for one batch) at step 600: 1.6326489448547363
Seen so far: 38464 sample
```

(3) 自定义训练和验证循环，代码如下。

```
#引入TensorFlow模块并基于TensorFlow构建Keras
```

```python
from __future__ import absolute_import, division, print_function
import tensorflow as tf
import tensorflow.keras as keras
import tensorflow.keras.layers as layers
#销毁当前的TF图并创建一个新图，该操作有助于避免旧模型或图层混乱
tf.keras.backend.clear_session()
#导入数据集并划分训练集和测试集
(x_train, y_train), (x_test, y_test) = keras.datasets.mnist.load_data()
x_train = x_train.reshape(60000, 784).astype('float32') /255
x_test = x_test.reshape(10000, 784).astype('float32') /255
x_val = x_train[-10000:]
y_val = y_train[-10000:]
x_train = x_train[:-10000]
y_train = y_train[:-10000]
#构建模型
inputs = keras.Input(shape=(784,), name='digits')
x = layers.Dense(64, activation='relu', name='dense_1')(inputs)
x = layers.Dense(64, activation='relu', name='dense_2')(x)
outputs = layers.Dense(10, activation='softmax', name='predictions')(x)
model = keras.Model(inputs=inputs, outputs=outputs)
#建立优化器
optimizer = keras.optimizers.SGD(learning_rate=1e-3)
#损失函数
loss_fn = keras.losses.SparseCategoricalCrossentropy()
#设定统计参数
train_acc_metric = keras.metrics.SparseCategoricalAccuracy()
val_acc_metric = keras.metrics.SparseCategoricalAccuracy()
#准备训练数据
batch_size = 64
train_dataset = tf.data.Dataset.from_tensor_slices((x_train, y_train))
train_dataset = train_dataset.shuffle(buffer_size=1024).batch(batch_size)
#准备验证数据
val_dataset = tf.data.Dataset.from_tensor_slices((x_val, y_val))
val_dataset = val_dataset.batch(64)
#构造循环
for epoch in range(3):
    print('Start of epoch %d' % (epoch,))
    for step, (x_batch_train, y_batch_train) in enumerate(train_dataset):
        with tf.GradientTape() as tape:
            logits = model(x_batch_train)
            loss_value = loss_fn(y_batch_train, logits)
        grads = tape.gradient(loss_value, model.trainable_variables)
        optimizer.apply_gradients(zip(grads, model.trainable_variables))
        #更新统计参数
```

```
            train_acc_metric(y_batch_train, logits)
        #输出
        if step % 200 == 0:
            print('Training loss (for one batch) at step %s: %s' % (step, float(loss_value)))
            print('Seen so far: %s samples' % ((step + 1) * 64))
    #输出统计参数的值
    train_acc = train_acc_metric.result()
    print('Training acc over epoch: %s' % (float(train_acc),))
    #重置统计参数
    train_acc_metric.reset_states()
    #验证
    for x_batch_val, y_batch_val in val_dataset:
        val_logits = model(x_batch_val)
        val_acc_metric(y_batch_val, val_logits)
    val_acc = val_acc_metric.result()
    val_acc_metric.reset_states()
    print('Validation acc: %s' % (float(val_acc),))
```

(4) 代码的运行结果如下。

```
Start of epoch 0
Training loss (for one batch) at step 0: 2.3243019580841064
Seen so far: 64 samples
...
Training loss (for one batch) at step 600: 2.0378403663635254
Seen so far: 38464 samples
Training acc over epoch: 0.24595999717712402
Validation acc: 0.4284000098705292
Start of epoch 1
Training loss (for one batch) at step 0: 1.9687241315841675
Seen so far: 64 samples
...
Training loss (for one batch) at step 600: 1.6812344789505005
Seen so far: 38464 samples
Training acc over epoch: 0.5197200179100037
Validation acc: 0.6035000085830688
Start of epoch 2
Training loss (for one batch) at step 0: 1.5821621417999268
Seen so far: 64 samples
...
Training loss (for one batch) at step 600: 1.2043852806091309
Seen so far: 38464 samples
Training acc over epoch: 0.6480000019073486
Validation acc: 0.7148000001907349
```

(5)可以在模型中添加自定义的loss,代码如下。

```
#引入TensorFlow模块并基于TensorFlow构建Keras
from __future__ import absolute_import, division, print_function
import tensorflow as tf
import tensorflow.keras as keras
import tensorflow.keras.layers as layers
#销毁当前的TF图并创建一个新图,该操作有助于避免旧模型或图层混乱
tf.keras.backend.clear_session()
#导入数据集并划分训练集和测试集
(x_train, y_train), (x_test, y_test) = keras.datasets.mnist.load_data()
x_train = x_train.reshape(60000, 784).astype('float32') /255
x_test = x_test.reshape(10000, 784).astype('float32') /255
x_val = x_train[-10000:]
y_val = y_train[-10000:]
x_train = x_train[:-10000]
y_train = y_train[:-10000]
#建立模型
inputs = keras.Input(shape=(784,), name='digits')
x = layers.Dense(64, activation='relu', name='dense_1')(inputs)
x = layers.Dense(64, activation='relu', name='dense_2')(x)
outputs = layers.Dense(10, activation='softmax', name='predictions')(x)
model = keras.Model(inputs=inputs, outputs=outputs)
#建立优化器
optimizer = keras.optimizers.SGD(learning_rate=1e-3)
#损失函数
loss_fn = keras.losses.SparseCategoricalCrossentropy()
#设定统计参数
train_acc_metric = keras.metrics.SparseCategoricalAccuracy()
val_acc_metric = keras.metrics.SparseCategoricalAccuracy()
#准备训练数据
batch_size = 64
train_dataset = tf.data.Dataset.from_tensor_slices((x_train, y_train))
train_dataset = train_dataset.shuffle(buffer_size=1024).batch(batch_size)
#准备验证数据
val_dataset = tf.data.Dataset.from_tensor_slices((x_val, y_val))
val_dataset = val_dataset.batch(64)
#添加自定义的loss,只能看到最近一次训练增加的loss
class ActivityRegularizationLayer(layers.Layer):
    def call(self, inputs):
        self.add_loss(1e-2 * tf.reduce_sum(inputs))
        return inputs
inputs = keras.Input(shape=(784,), name='digits')
x = layers.Dense(64, activation='relu', name='dense_1')(inputs)
```

```python
#插入激活层
x = ActivityRegularizationLayer()(x)
x = layers.Dense(64, activation='relu', name='dense_2')(x)
outputs = layers.Dense(10, activation='softmax', name='predictions')(x)
#构建模型
model = keras.Model(inputs=inputs, outputs=outputs)
logits = model(x_train[:64])
print(model.losses)
logits = model(x_train[:64])
logits = model(x_train[64: 128])
logits = model(x_train[128: 192])
print(model.losses)
#将loss添加到求导中
optimizer = keras.optimizers.SGD(learning_rate=1e-3)
for epoch in range(3):
    print('Start of epoch %d' % (epoch,))
    for step, (x_batch_train, y_batch_train) in enumerate(train_dataset):
        with tf.GradientTape() as tape:
            logits = model(x_batch_train)
            loss_value = loss_fn(y_batch_train, logits)
            #添加额外的loss
            loss_value += sum(model.losses)
        grads = tape.gradient(loss_value, model.trainable_variables)
        optimizer.apply_gradients(zip(grads, model.trainable_variables))
        #将200个样本做为1个batch并输出一次学习日志
        if step % 200 == 0:
            print('Training loss (for one batch) at step %s:
                %s' % (step, float(loss_value)))
            print('Seen so far: %s samples' % ((step + 1) * 64))
```

（6）代码的运行结果如下。

```
Start of epoch 0
raining loss (for one batch) at step 0: 8.432535171508789
Seen so far: 64 samples
...
Training loss (for one batch) at step 600: 2.3722951412200928
Seen so far: 38464 samples
Start of epoch 1
Training loss (for one batch) at step 0: 2.3349554538726807
Seen so far: 64 samples
...
Training loss (for one batch) at step 600: 2.3314294815063477
Seen so far: 38464 samples
Start of epoch 2
```

```
Training loss (for one batch) at step 0: 2.3312125205993652
Seen so far: 64 samples
...
Training loss (for one batch) at step 600: 2.308372974395752
Seen so far: 38464 samples
```

7.5　Keras 模型的保存

因为深度学习模型的训练时长可能为数小时、数天甚至数周，所以有必要在训练的过程中对模型进行保存，本节介绍如何将 Keras 模型保存到文件中并再次加载。

（1）构建一个简单的网络模型，代码如下。

```
#引入 TensorFlow 模块并基于 TensorFlow 构建 Keras
from __future__ import absolute_import, division, print_function
import tensorflow as tf
import tensorflow.keras as keras
import tensorflow.keras.layers as layers
#销毁当前的 TF 图并创建一个新图，该操作有助于避免旧模型或图层混乱
tf.keras.backend.clear_session()
#定义 Keras 张量，并将其作为输入
inputs = keras.Input(shape=(784,), name='digits')
#构建网络层
x = layers.Dense(64, activation='relu', name='dense_1')(inputs)
x = layers.Dense(64, activation='relu', name='dense_2')(x)
outputs = layers.Dense(10, activation='softmax', name='predictions')(x)
#构建模型
model = keras.Model(inputs=inputs, outputs=outputs, name='7_5_test')
#输出模型详情
model.summary()
#导入数据集并划分训练集和测试集
(x_train, y_train), (x_test, y_test) = keras.datasets.mnist.load_data()
x_train = x_train.reshape(60000, 784).astype('float32') / 255
x_test = x_test.reshape(10000, 784).astype('float32') / 255
#初始化模型参数
model.compile(loss='sparse_categorical_crossentropy',
    optimizer=keras.optimizers.RMSprop())
#训练模型
history = model.fit(x_train, y_train, batch_size=64, epochs=1)
#评估模型
predictions = model.predict(x_test)
```

（2）代码的运行结果如下。

```
Model: "7_5_test"
_____
```

```
Layer (type)                 Output Shape              Param #
=================================================================
digits (InputLayer)          [(None, 784)]             0
_____
dense_1 (Dense)              (None, 64)                50240
_____
dense_2 (Dense)              (None, 64)                4160
_____
predictions (Dense)          (None, 10)                650
=================================================================
Total params: 55,050
Trainable params: 55,050
Non-trainable params: 0
_____
Train on 60000 samples
60000/60000 [==============================] - 3s 55us/sample - loss: 0.3117
```

（3）可以保存整个模型，保存的内容包括模型的架构、模型的权重、模型的训练配置、优化器及其状态。保存上面构建的简单网络，读取后进行验证，代码如下。

```
#引入TensorFlow模块并基于TensorFlow构建Keras
from __future__ import absolute_import, division, print_function
import tensorflow as tf
import tensorflow.keras as keras
import tensorflow.keras.layers as layers
#销毁当前的TF图并创建一个新图，该操作有助于避免旧模型或图层混乱
tf.keras.backend.clear_session()
#定义Keras张量，并将其作为输入
inputs = keras.Input(shape=(784,), name='digits')
#构建网络层
x = layers.Dense(64, activation='relu', name='dense_1')(inputs)
x = layers.Dense(64, activation='relu', name='dense_2')(x)
outputs = layers.Dense(10, activation='softmax', name='predictions')(x)
#构建模型
model = keras.Model(inputs=inputs, outputs=outputs, name='7_5_test')
#输出模型详情
model.summary()
#导入数据集并划分训练集和测试集
(x_train, y_train), (x_test, y_test) = keras.datasets.mnist.load_data()
x_train = x_train.reshape(60000, 784).astype('float32') / 255
x_test = x_test.reshape(10000, 784).astype('float32') / 255
#初始化模型参数
model.compile(loss='sparse_categorical_crossentropy',
    optimizer=keras.optimizers.RMSprop())
#训练模型
```

```
history = model.fit(x_train, y_train, batch_size=64, epochs=1)
#评估模型
predictions = model.predict(x_test)
#输出模型评估结果
print (predictions)
#导入NumPy模块
import numpy as np
#将模型保存为the_save_model.h5文件
model.save('the_save_model.h5')
#从保存的the_save_model.h5文件中载入模型
new_model = keras.models.load_model('the_save_model.h5')
#对载入的模型进行评估
new_prediction = new_model.predict(x_test)
#输出评估结果
print (new_prediction)
#如果结果不一致则抛出异常，反之忽略
np.testing.assert_allclose(predictions, new_prediction, atol=1e-6)
```

说明：NumPy是Python的扩展库，提供了大量基于数组的计算函数。

（4）代码的运行结果如下。

```
Model: "7_5_test"
_____
Layer (type)                 Output Shape              Param #
=================================================================
digits (InputLayer)          [(None, 784)]             0
_____
dense_1 (Dense)              (None, 64)                50240
_____
dense_2 (Dense)              (None, 64)                4160
_____
predictions (Dense)          (None, 10)                650
=================================================================
Total params: 55,050
Trainable params: 55,050
Non-trainable params: 0
_____
Train on 60000 samples
60000/60000 [==============================] - 3s 51us/sample - loss: 0.3091
[[2.92701725e-06 2.99165276e-06 5.59028122e-05 ... 9.93639290e-01
  1.23887032e-04 2.07889636e-04]
 [2.84634414e-04 1.35285882e-04 9.82001007e-01 ... 3.05208125e-09
  7.95908636e-05 2.72367533e-08]
 [2.70078272e-05 9.84026670e-01 2.21077469e-03 ... 1.07535403e-02
  3.86661210e-04 1.92594511e-04]
```

```
    ...
    [3.12386419e-07 4.55297631e-08 7.52926690e-06 ... 7.87596276e-04
     6.85557025e-05 4.04629204e-03]
    [9.28272527e-07 7.03017167e-06 5.38014788e-08 ... 2.07835157e-07
     4.47527040e-04 2.50885887e-06]
    [1.42830038e-06 9.81979428e-11 3.56303417e-06 ... 1.45473494e-11
     4.37053815e-09 1.37566722e-10]]


[[2.92701725e-06 2.99165276e-06 5.59028122e-05 ... 9.93639290e-01
  1.23887032e-04 2.07889636e-04]
 [2.84634414e-04 1.35285882e-04 9.82001007e-01 ... 3.05208125e-09
  7.95908636e-05 2.72367533e-08]
 [2.70078272e-05 9.84026670e-01 2.21077469e-03 ... 1.07535403e-02
  3.86661210e-04 1.92594511e-04]
 ...
 [3.12386419e-07 4.55297631e-08 7.52926690e-06 ... 7.87596276e-04
  6.85557025e-05 4.04629204e-03]
 [9.28272527e-07 7.03017167e-06 5.38014788e-08 ... 2.07835157e-07
  4.47527040e-04 2.50885887e-06]
 [1.42830038e-06 9.81979428e-11 3.56303417e-06 ... 1.45473494e-11
  4.37053815e-09 1.37566722e-10]]
```

输出结果没有抛出异常，在保存和读取模型的过程中，没有出现预测结果不一致的情况。在相应的目录下会生成指定文件名的文件，本例中生成的文件是 the_save_model.h5。

（5）可以将模型保存为 SavedModel 文件，代码如下。

```
#引入TensorFlow模块并基于TensorFlow构建Keras
from __future__ import absolute_import, division, print_function
import tensorflow as tf
import tensorflow.keras as keras
import tensorflow.keras.layers as layers
#销毁当前的TF图并创建一个新图,该操作有助于避免旧模型或图层混乱
tf.keras.backend.clear_session()
#定义Keras张量,并将其作为输入
inputs = keras.Input(shape=(784,), name='digits')
#构建网络层
x = layers.Dense(64, activation='relu', name='dense_1')(inputs)
x = layers.Dense(64, activation='relu', name='dense_2')(x)
outputs = layers.Dense(10, activation='softmax', name='predictions')(x)
#构建模型
model = keras.Model(inputs=inputs, outputs=outputs, name='7_5_test')
#输出模型详情
model.summary()
#导入数据集并划分训练集和测试集
```

```python
(x_train, y_train), (x_test, y_test) = keras.datasets.mnist.load_data()
x_train = x_train.reshape(60000, 784).astype('float32') / 255
x_test = x_test.reshape(10000, 784).astype('float32') / 255
#初始化模型参数
model.compile(loss='sparse_categorical_crossentropy',
    optimizer=keras.optimizers.RMSprop())
#训练模型
history = model.fit(x_train, y_train, batch_size=64, epochs=1)
#评估模型
predictions = model.predict(x_test)
#输出模型评估结果
print (predictions)
#导入 NumPy 模块
import numpy as np
#将模型保存为 SavedModel 文件
keras.experimental.export_saved_model(model, 'saved_model')
#从保存的文件中载入模型
new_model = keras.experimental.load_from_saved_model('saved_model')
#对载入的模型进行评估
new_prediction = new_model.predict(x_test)
#如果结果不一致则抛出异常，反之忽略
np.testing.assert_allclose(predictions, new_prediction, atol=1e-6)
```

（6）代码的运行结果如下。

```
Model: "7_5_test"
_____
Layer (type)                 Output Shape              Param #
=================================================================
digits (InputLayer)          [(None, 784)]             0
_____
dense_1 (Dense)              (None, 64)                50240
_____
dense_2 (Dense)              (None, 64)                4160
_____
predictions (Dense)          (None, 10)                650
=================================================================
Total params: 55,050
Trainable params: 55,050
Non-trainable params: 0
_____
Train on 60000 samples
60000/60000 [==============================] - 3s 55us/sample - loss: 0.3152
```

输出结果没有抛出异常，在保存和读取模型的过程中，没有出现预测结果不一致的情况。在相应的目录下会生成指定文件名的文件，本例中生成的文件是 save_model。

（7）在实际项目中，可以仅保存网络结构。在这种情况下，导出的模型中不包含训练好的参数，代码如下。

```python
#引入 TensorFlow 模块并基于 TensorFlow 构建 Keras
from __future__ import absolute_import, division, print_function
import tensorflow as tf
import tensorflow.keras as keras
import tensorflow.keras.layers as layers
#销毁当前的 TF 图并创建一个新图，该操作有助于避免旧模型或图层混乱
tf.keras.backend.clear_session()
#定义 Keras 张量，并将其作为输入
inputs = keras.Input(shape=(784,), name='digits')
#构建网络层
x = layers.Dense(64, activation='relu', name='dense_1')(inputs)
x = layers.Dense(64, activation='relu', name='dense_2')(x)
outputs = layers.Dense(10, activation='softmax', name='predictions')(x)
#构建模型
model = keras.Model(inputs=inputs, outputs=outputs, name='7_5_test')
#输出模型详情
model.summary()
#导入数据集并划分训练集和测试集
(x_train, y_train), (x_test, y_test) = keras.datasets.mnist.load_data()
x_train = x_train.reshape(60000, 784).astype('float32') / 255
x_test = x_test.reshape(10000, 784).astype('float32') / 255
#初始化模型参数
model.compile(loss='sparse_categorical_crossentropy',
    optimizer=keras.optimizers.RMSprop())
#训练模型
history = model.fit(x_train, y_train, batch_size=64, epochs=1)
#评估模型
predictions = model.predict(x_test)
#导入 NumPy 模块
import numpy as np
#返回包含模型配置信息的 Python 字典。模型也可以从它的 config 信息中重构回去
config = model.get_config()
#根据保存的模型配置信息构建新模型
reinitialized_model = keras.Model.from_config(config)
#对新模型进行评估
new_prediction = reinitialized_model.predict(x_test)
#判断评估结果是否一致
if abs(np.sum(predictions-new_prediction)) > 0 :
    print ("is same!")
else:
    print ("not same!")
```

(8) 代码的运行结果如下。

```
Model: "7_5_test"
_____
Layer (type)                 Output Shape              Param #
=================================================================
digits (InputLayer)          [(None, 784)]             0
_____
dense_1 (Dense)              (None, 64)                50240
_____
dense_2 (Dense)              (None, 64)                4160
_____
predictions (Dense)          (None, 10)                650
=================================================================
Total params: 55,050
Trainable params: 55,050
Non-trainable params: 0
_____
Train on 60000 samples
60000/60000 [==============================] - 4s 59us/sample - loss: 0.2959
is same!
```

(9) 在实际项目中,也可以使用 json 保存网络结构,代码如下。

```
#引入 TensorFlow 模块并基于 TensorFlow 构建 Keras
from __future__ import absolute_import, division, print_function
import tensorflow as tf
import tensorflow.keras as keras
import tensorflow.keras.layers as layers
#销毁当前的 TF 图并创建一个新图,该操作有助于避免旧模型或图层混乱
tf.keras.backend.clear_session()
#定义 Keras 张量,并将其作为输入
inputs = keras.Input(shape=(784,), name='digits')
#构建网络层
x = layers.Dense(64, activation='relu', name='dense_1')(inputs)
x = layers.Dense(64, activation='relu', name='dense_2')(x)
outputs = layers.Dense(10, activation='softmax', name='predictions')(x)
#构建模型
model = keras.Model(inputs=inputs, outputs=outputs, name='7_5_test')
#输出模型详情
model.summary()
#导入数据集并划分训练集和测试集
(x_train, y_train), (x_test, y_test) = keras.datasets.mnist.load_data()
x_train = x_train.reshape(60000, 784).astype('float32') / 255
x_test = x_test.reshape(10000, 784).astype('float32') / 255
#初始化模型参数
```

```
model.compile(loss='sparse_categorical_crossentropy',
    optimizer=keras.optimizers.RMSprop())
#训练模型
history = model.fit(x_train, y_train, batch_size=64, epochs=1)
#评估模型
predictions = model.predict(x_test)
#导入 NumPy 模块
import numpy as np
#将模型的配置保存为 json 格式
json_config = model.to_json()
#使用 json 构建新模型
reinitialized_model = keras.models.model_from_json(json_config)
#对新模型进行评估
new_prediction = reinitialized_model.predict(x_test)
#判断评估结果是否一致
if abs(np.sum(predictions-new_prediction)) > 0 :
    print ( " is same! " )
else:
    print ( " not same! " )
```

（10）代码的运行结果如下。

```
Model: " 7_5_test "
_____
Layer (type)                 Output Shape              Param #
=================================================================
digits (InputLayer)          [(None, 784)]             0
_____
dense_1 (Dense)              (None, 64)                50240
_____
dense_2 (Dense)              (None, 64)                4160
_____
predictions (Dense)          (None, 10)                650
=================================================================
Total params: 55,050
Trainable params: 55,050
Non-trainable params: 0
_____
Train on 60000 samples
60000/60000 [==============================] - 4s 59us/sample - loss: 0.2959
is same!
```

（11）在实际项目中，也可以仅保存网络参数，代码如下。

```
#引入 TensorFlow 模块并基于 TensorFlow 构建 Keras
from __future__ import absolute_import, division, print_function
import tensorflow as tf
```

```python
import tensorflow.keras as keras
import tensorflow.keras.layers as layers
#销毁当前的 TF 图并创建一个新图，该操作有助于避免旧模型或图层混乱
tf.keras.backend.clear_session()
#定义 Keras 张量，并将其作为输入
inputs = keras.Input(shape=(784,), name='digits')
#构建网络层
x = layers.Dense(64, activation='relu', name='dense_1')(inputs)
x = layers.Dense(64, activation='relu', name='dense_2')(x)
outputs = layers.Dense(10, activation='softmax', name='predictions')(x)
#构建模型
model = keras.Model(inputs=inputs, outputs=outputs, name='7_5_test')
#输出模型详情
model.summary()
#导入数据集并划分训练集和测试集
(x_train, y_train), (x_test, y_test) = keras.datasets.mnist.load_data()
x_train = x_train.reshape(60000, 784).astype('float32') / 255
x_test = x_test.reshape(10000, 784).astype('float32') / 255
#初始化模型参数
model.compile(loss='sparse_categorical_crossentropy',
    optimizer=keras.optimizers.RMSprop())
#训练模型
history = model.fit(x_train, y_train, batch_size=64, epochs=1)
#评估模型
predictions = model.predict(x_test)
#导入 NumPy 模块
import numpy as np
#获取该层的参数 w 和 b
weights = model.get_weights()
model.set_weights(weights)
#保存参数
config = model.get_config()
weights = model.get_weights()
new_model = keras.Model.from_config(config)
#config 只能使用 keras.Model 的 API
new_model.set_weights(weights)
#获得新模型的评估结果
new_predictions = new_model.predict(x_test)
#比较评估结果
np.testing.assert_allclose(predictions, new_predictions, atol=1e-6)
```

（12）代码的运行结果如下。

```
Model: "7_5_test"
_____
```

```
Layer (type)                 Output Shape              Param #
=================================================================
digits (InputLayer)          [(None, 784)]             0
_____
dense_1 (Dense)              (None, 64)                50240
_____
dense_2 (Dense)              (None, 64)                4160
_____
predictions (Dense)          (None, 10)                650
=================================================================
Total params: 55,050
Trainable params: 55,050
Non-trainable params: 0
_____
Train on 60000 samples
60000/60000 [==============================] - 3s 49us/sample - loss: 0.3087
```

没有报出异常，说明相关模型的保存和提取没有问题。

（13）在实际项目中，常常需要保存子类模型的参数，无法保存和序列化子类模型的结构，只能保存参数。构建模型并训练，代码如下。

```
#引入TensorFlow模块并基于TensorFlow构建Keras
from __future__ import absolute_import, division, print_function
import tensorflow as tf
import tensorflow.keras as keras
import tensorflow.keras.layers as layers
#销毁当前的TF图并创建一个新图，该操作有助于避免旧模型或图层混乱
tf.keras.backend.clear_session()
#导入NumPy模块
import numpy as np
#定义模型类
class ThreeLayerMLP(keras.Model):
    def __init__(self, name=None):
        super(ThreeLayerMLP, self).__init__(name=name)
        self.dense_1 = layers.Dense(64, activation='relu', name='dense_1')
        self.dense_2 = layers.Dense(64, activation='relu', name='dense_2')
        self.pred_layer = layers.Dense(10, activation='softmax', name='predictions')
    def call(self, inputs):
        x = self.dense_1(inputs)
        x = self.dense_2(x)
        return self.pred_layer(x)
def get_model():
    return ThreeLayerMLP(name='3_layer_mlp')
#使用定义的模型类构建模型
model = get_model()
```

```python
#导入数据集并划分训练集和测试集
(x_train, y_train), (x_test, y_test) = keras.datasets.mnist.load_data()
x_train = x_train.reshape(60000, 784).astype('float32') / 255
x_test = x_test.reshape(10000, 784).astype('float32') / 255
#初始化模型参数
model.compile(loss='sparse_categorical_crossentropy',
    optimizer=keras.optimizers.RMSprop())
#训练模型
history = model.fit(x_train, y_train, batch_size=64,epochs=1)
#保存权重参数
model.save_weights('my_model_weights', save_format='tf')
#输出结果
predictions = model.predict(x_test)
first_batch_loss = model.train_on_batch(x_train[:64], y_train[:64])
#读取保存的模型参数
new_model = get_model()
new_model.compile(loss='sparse_categorical_crossentropy',
    optimizer=keras.optimizers.RMSprop())
#根据保存的模型参数构建新模型
new_model.load_weights('my_model_weights')
#对新模型进行评估
new_predictions = new_model.predict(x_test)
np.testing.assert_allclose(predictions, new_predictions, atol=1e-6)
new_first_batch_loss = new_model.train_on_batch(x_train[:64], y_train[:64])
#判断评估结果是否一致
if first_batch_loss == new_first_batch_loss:
    print ( " is same! " )
else:
    print( " not same! " )
```

（14）代码的运行结果如下。

```
Train on 60000 samples
60000/60000 [==============================] - 3s 50us/sample - loss: 0.3171
is same!
```

（15）为了恢复模型并进行训练或评估，可以设置相应的检查点文件，代码如下。

```python
#引入 TensorFlow 模块并基于 TensorFlow 构建 Keras
from __future__ import absolute_import, division, print_function
import tensorflow as tf
import tensorflow.keras as keras
import tensorflow.keras.layers as layers
#销毁当前的 TF 图并创建一个新图，该操作有助于避免旧模型或图层混乱
tf.keras.backend.clear_session()
#导入 NumPy 模块
import numpy as np
```

```python
import os
#定义模型类
class ThreeLayerMLP(keras.Model):
    def __init__(self, name=None):
        super(ThreeLayerMLP, self).__init__(name=name)
        self.dense_1 = layers.Dense(64, activation='relu', name='dense_1')
        self.dense_2 = layers.Dense(64, activation='relu', name='dense_2')
        self.pred_layer = layers.Dense(10, activation='softmax', name='predictions')
    def call(self, inputs):
        x = self.dense_1(inputs)
        x = self.dense_2(x)
        return self.pred_layer(x)
def get_model():
    return ThreeLayerMLP(name='3_layer_mlp')
#定义检查点
check_path = 'model.ckpt'
check_dir = os.path.dirname(check_path)
#定义回调函数
cp_callback = tf.keras.callbacks.ModelCheckpoint(check_path,
    save_weights_only=True, verbose=1)
#使用定义的模型类构建模型
model = get_model()
#导入数据集并划分训练集和测试集
(x_train, y_train), (x_test, y_test) = keras.datasets.mnist.load_data()
x_train = x_train.reshape(60000, 784).astype('float32') / 255
x_test = x_test.reshape(10000, 784).astype('float32') / 255
#初始化模型参数
model.compile(loss='sparse_categorical_crossentropy',
    optimizer=keras.optimizers.RMSprop())
#训练模型
history = model.fit(x_train, y_train, batch_size=64,
    epochs=1, callbacks=[cp_callback])
```

（16）代码的运行结果如下。

```
Train on 60000 samples
59968/60000 [============================>.] - ETA: 0s - loss: 0.3116
Epoch 00001: saving model to model.ckpt
60000/60000 [==============================] - 4s 64us/sample - loss: 0.3115
```

（17）在实际项目中，需要按照一定规则生成检查点文件，代码如下。

```
#引入 TensorFlow 模块并基于 TensorFlow 构建 Keras
from __future__ import absolute_import, division, print_function
import tensorflow as tf
import tensorflow.keras as keras
import tensorflow.keras.layers as layers
```

```
#销毁当前的TF图并创建一个新图,该操作有助于避免旧模型或图层混乱
tf.keras.backend.clear_session()
#导入NumPy模块
import numpy as np
import os
#定义模型类
class ThreeLayerMLP(keras.Model):
    def __init__(self, name=None):
        super(ThreeLayerMLP, self).__init__(name=name)
        self.dense_1 = layers.Dense(64, activation='relu', name='dense_1')
        self.dense_2 = layers.Dense(64, activation='relu', name='dense_2')
        self.pred_layer = layers.Dense(10, activation='softmax', name='predictions')
    def call(self, inputs):
        x = self.dense_1(inputs)
        x = self.dense_2(x)
        return self.pred_layer(x)
def get_model():
    return ThreeLayerMLP(name='3_layer_mlp')
#定义检查点
check_path = 'model.ckpt'
check_dir = os.path.dirname(check_path)
#定义回调函数,每训练5次回调一次,生成一个检查点文件
cp_callback = tf.keras.callbacks.ModelCheckpoint(check_path,
    save_weights_only=True, verbose=1, period=5)
#使用定义的模型类构建模型
model = get_model()
#导入数据集并划分训练集和测试集
(x_train, y_train), (x_test, y_test) = keras.datasets.mnist.load_data()
x_train = x_train.reshape(60000, 784).astype('float32') / 255
x_test = x_test.reshape(10000, 784).astype('float32') / 255
#初始化模型参数
model.compile(loss='sparse_categorical_crossentropy',
    optimizer=keras.optimizers.RMSprop())
#训练模型
history = model.fit(x_train, y_train, batch_size=64,
    epochs=1, callbacks=[cp_callback])
```

(18)代码的运行结果如下,可以看出,每训练5次生成一个检查点文件。

```
Train on 60000 samples
Epoch 1/10
60000/60000 [==============================] - 3s 48us/sample - loss: 0.3216
...
Epoch 5/10
59264/60000 [=============================>.] - ETA: 0s - loss: 0.0707
```

```
Epoch 00005: saving model to cp-0005.ckpt
...
Epoch 10/10
59904/60000 [============================>.] - ETA: 0s - loss: 0.0376
Epoch 00010: saving model to cp-0010.ckpt
60000/60000 [=============================] - 2s 38us/sample - loss: 0.0376
```

（19）载入最新的检查点文件，代码如下。

```
#引入 TensorFlow 模块并基于 TensorFlow 构建 Keras
from __future__ import absolute_import, division, print_function
import tensorflow as tf
import tensorflow.keras as keras
import tensorflow.keras.layers as layers
#销毁当前的 TF 图并创建一个新图，该操作有助于避免旧模型或图层混乱
tf.keras.backend.clear_session()
#导入 NumPy 模块
import numpy as np
import os
#定义模型类
class ThreeLayerMLP(keras.Model):
    def __init__(self, name=None):
        super(ThreeLayerMLP, self).__init__(name=name)
        self.dense_1 = layers.Dense(64, activation='relu', name='dense_1')
        self.dense_2 = layers.Dense(64, activation='relu', name='dense_2')
        self.pred_layer = layers.Dense(10, activation='softmax', name='predictions')
    def call(self, inputs):
        x = self.dense_1(inputs)
        x = self.dense_2(x)
        return self.pred_layer(x)
def get_model():
    return ThreeLayerMLP(name='3_layer_mlp')
#定义检查点
check_path = 'model.ckpt'
check_dir = os.path.dirname(check_path)
#定义回调函数，每训练 5 次回调一次，生成一个检查点文件
cp_callback = tf.keras.callbacks.ModelCheckpoint(check_path,
    save_weights_only=True, verbose=1,period=5)
#导入数据集并划分训练集和测试集
(x_train, y_train), (x_test, y_test) = keras.datasets.mnist.load_data()
x_train = x_train.reshape(60000, 784).astype('float32') / 255
x_test = x_test.reshape(10000, 784).astype('float32') / 255
#载入最新的检查点文件
latest = tf.train.latest_checkpoint(check_dir)
print(latest)
```

```
#构建模型
model = get_model()
#从最新的检查点载入新模型的配置
model.load_weights(latest)
#初始化模型参数
model.compile(loss='sparse_categorical_crossentropy',
    optimizer=keras.optimizers.RMSprop())
#训练模型
history = model.fit(x_train, y_train, batch_size=64,
    epochs=10, callbacks=[cp_callback])
```

（20）代码的运行结果如下。

```
cp-0010.ckpt
Train on 60000 samples
Epoch 1/10
60000/60000 [==============================] - 3s 56us/sample - loss: 2.6940e-04
...
Epoch 5/10
59648/60000 [==========================>.] - ETA: 0s - loss: 4.6151e-04
Epoch 00005: saving model to cp-0005.ckpt
60000/60000 [==============================] - 3s 43us/sample - loss: 4.5880e-04
...
Epoch 10/10
58880/60000 [=========================>.] - ETA: 0s - loss: 4.3697e-04
Epoch 00010: saving model to cp-0010.ckpt
60000/60000 [==============================] - 3s 44us/sample - loss: 4.2882e-04
```

本例中提取最新的检查点进行了训练，并记录检查点。使用这种方法可以对训练时间较长的模型进行阶段性保存，使结果更加准确。

Keras 是一个非常方便的深度学习框架。使用 Keras 可以快速搭建深度学习网络、灵活地选取训练参数、降低机器学习编码的使用难度。

第 8 章 TensorFlow 文本分类

从本章开始,将通过介绍实际的例子对前面学习的知识进行阶段性回顾。本章对文本分类进行介绍。

8.1 简单文本分类

文本分类一直是统计学中的重要课题,应用机器学习技术能够提高文本分类的速度和准确率。本节对 TensorFlow 2.0 在文本分类领域的应用进行初步探讨。

(1)下载 IMDB 数据集并了解其特征,代码如下。

```
import ssl
#引入 TensorFlow 模块并基于 TensorFlow 构建 Keras
import tensorflow as tf
import tensorflow.keras as keras
import tensorflow.keras.layers as layers
ssl._create_default_https_context = ssl._create_unverified_context
#导入数据集并划分训练集和测试集
imdb=keras.datasets.imdb
(train_x, train_y), (test_x, text_y)=keras.datasets.imdb.load_data(num_words=10000)
#格式化输出训练集
print("Training entries: {}, labels: {}".format(len(train_x), len(train_y)))
print(train_x[0])
print('len: ',len(train_x[0]), len(train_x[1]))
```

(2)代码的运行结果如下。

```
Downloading data from https://storage.googleapis.com/tensorflow/tf-keras-datasets/imdb.npz
17465344/17464789 [==============================] - 23s 1us/step
Training entries: 25000, labels: 25000
[1, 14, 22, 16, 43, 530, 973, 1622, 1385, 65, 458, 4468, 66, 3941, 4, 173, 36, 256, 5, 25, 100, 43, 838, 112, 50, 670, 2, 9, 35, 480, 284, 5, 150, 4, 172, 112, 167, 2, 336, 385, 39, 4, 172, 4536, 1111, 17, 546, 38, 13, 447, 4, 192, 50, 16, 6, 147, 2025, 19, 14, 22, 4, 1920, 4613, 469, 4, 22, 71, 87, 12, 16, 43, 530, 38, 76, 15, 13, 1247, 4, 22, 17, 515, 17, 12, 16, 626, 18, 2, 5, 62, 386, 12, 8, 316, 8, 106, 5, 4, 2223, 5244, 16, 480, 66, 3785, 33, 4, 130, 12, 16, 38, 619, 5, 25, 124, 51, 36, 135, 48, 25, 1415, 33, 6, 22, 12, 215, 28, 77, 52, 5, 14, 407, 16, 82, 2, 8, 4, 107, 117, 5952, 15, 256, 4, 2, 7, 3766, 5, 723, 36, 71, 43, 530, 476, 26, 400, 317, 46, 7, 4, 2, 1029, 13, 104, 88, 4, 381, 15, 297, 98, 32, 2071, 56, 26, 141, 6, 194, 7486, 18, 4, 226, 22, 21, 134, 476, 26, 480, 5, 144, 30, 5535, 18, 51, 36, 28, 224, 92, 25, 104, 4, 226, 65, 16, 38, 1334, 88, 12, 16, 283, 5, 16, 4472, 113, 103, 32, 15, 16, 5345, 19, 178, 32]
len: 218 189
```

（3）在保证数据导入结果正确的情况下，文本分类需要创建 ID 和词的匹配字典，代码如下。

```
import ssl
#引入 TensorFlow 模块并基于 TensorFlow 构建 Keras
import tensorflow as tf
import tensorflow.keras as keras
import tensorflow.keras.layers as layers
ssl._create_default_https_context = ssl._create_unverified_context
#导入数据集并划分训练集和测试集
imdb=keras.datasets.imdb
(train_x, train_y), (test_x, text_y)=keras.datasets.imdb.load_data(num_words=10000)
#将数据转换成字典
word_index = imdb.get_word_index()
#将字典中的数据反转
word2id = {k:(v+3) for k, v in word_index.items()}
word2id['<PAD>'] = 0
word2id['<START>'] = 1
word2id['<UNK>'] = 2
word2id['<UNUSED>'] = 3
id2word = {v:k for k, v in word2id.items()}
def get_words(sent_ids):
    return ' '.join([id2word.get(i, '?') for i in sent_ids])
sent = get_words(train_x[0])
print(sent)
```

（4）代码的运行结果如下。

```
Downloading data from https://storage.googleapis.com/tensorflow/tf-keras-datasets/imdb_word_index.json
1646592/1641221 [==============================] - 1s 0us/step
<START> this film was just brilliant casting location scenery story direction everyone's really suited the part they played and you could just imagine being there robert <UNK> is an amazing actor and now the same being director <UNK> father came from the same scottish island as myself so i loved the fact there was a real connection with this film the witty remarks throughout the film were great it was just brilliant so much that i bought the film as soon as it was released for <UNK> and would recommend it to everyone to watch and the fly fishing was amazing really cried at the end it was so sad and you know what they say if you cry at a film it must have been good and this definitely was also <UNK> to the two little boy's that played the <UNK> of norman and paul they were just brilliant children are often left out of the <UNK> list i think because the stars that play them all grown up are such a big profile for the whole film but these children are amazing and should be praised for what they have done don't you think the whole story was so lovely because it was true and was someone's life after all that was shared with us all
```

（5）准备需要分类的数据，代码如下。

```
import ssl
#引入 TensorFlow 模块并基于 TensorFlow 构建 Keras
```

```python
import tensorflow as tf
import tensorflow.keras as keras
import tensorflow.keras.layers as layers
ssl._create_default_https_context = ssl._create_unverified_context
#导入数据集并划分训练集和测试集
imdb=keras.datasets.imdb
(train_x, train_y), (test_x, text_y)=keras.datasets.imdb.load_data(num_words=10000)
#将数据转换成字典
word_index = imdb.get_word_index()
word2id = {k:(v+3) for k, v in word_index.items()}
word2id['<PAD>'] = 0
word2id['<START>'] = 1
word2id['<UNK>'] = 2
word2id['<UNUSED>'] = 3
id2word = {v:k for k, v in word2id.items()}
def get_words(sent_ids):
    return ' '.join([id2word.get(i, '?') for i in sent_ids])
sent = get_words(train_x[0])
#句子末尾padding
train_x = keras.preprocessing.sequence.pad_sequences(
    train_x, value=word2id['<PAD>'],
    padding='post', maxlen=256
)
test_x = keras.preprocessing.sequence.pad_sequences(
    test_x, value=word2id['<PAD>'],
    padding='post', maxlen=256
)
print(train_x[0])
print('len: ',len(train_x[0]), len(train_x[1]))
```

（6）代码的运行结果如下。

```
[   1   14   22   16   43  530  973 1622 1385   65  458 4468   66 3941
    4  173   36  256    5   25  100   43  838  112   50  670    2    9
   35  480  284    5  150    4  172  112  167    2  336  385   39    4
  172 4536 1111   17  546   38   13  447    4  192   50   16    6  147
 2025   19   14   22    4 1920 4613  469    4   22   71   87   12   16
   43  530   38   76   15   13 1247    4   22   17  515   17   12   16
  626   18    2    5   62  386   12    8  316    8  106    5    4 2223
 5244   16  480   66 3785   33    4  130   12   16   38  619    5   25
  124   51   36  135   48   25 1415   33    6   22   12  215   28   77
   52    5   14  407   16   82    2    8    4  107  117 5952   15  256
    4    2    7 3766    5  723   36   71   43  530  476   26  400  317
   46    7    4    2 1029   13  104   88    4  381   15  297   98   32
 2071   56   26  141    6  194 7486   18    4  226   22   21  134  476
```

```
      26  480    5  144   30 5535   18   51   36   28  224   92   25  104
       4  226   65   16   38 1334   88   12   16  283    5   16 4472  113
     103   32   15   16 5345   19  178   32    0    0    0    0    0    0
       0    0    0    0    0    0    0    0    0    0    0    0    0    0
       0    0    0    0    0    0    0    0    0    0    0    0    0    0
       0    0    0    0]
len:  256 256
```

（7）构建文本分类模型，代码如下。

```
import ssl
#引入TensorFlow模块并基于TensorFlow构建Keras
import tensorflow as tf
import tensorflow.keras as keras
import tensorflow.keras.layers as layers
ssl._create_default_https_context = ssl._create_unverified_context
#导入数据集并划分训练集和测试集
imdb=keras.datasets.imdb
(train_x, train_y), (test_x, text_y)=keras.datasets.imdb.load_data(num_words=10000)
#将数据转换成字典
word_index = imdb.get_word_index()
word2id = {k:(v+3) for k, v in word_index.items()}
word2id['<PAD>'] = 0
word2id['<START>'] = 1
word2id['<UNK>'] = 2
word2id['<UNUSED>'] = 3
id2word = {v:k for k, v in word2id.items()}
def get_words(sent_ids):
    return ' '.join([id2word.get(i, '?') for i in sent_ids])
sent = get_words(train_x[0])
#句子末尾padding
train_x = keras.preprocessing.sequence.pad_sequences(
    train_x, value=word2id['<PAD>'],
    padding='post', maxlen=256
)
test_x = keras.preprocessing.sequence.pad_sequences(
    test_x, value=word2id['<PAD>'],
    padding='post', maxlen=256
)
#定义词汇size
vocab_size = 10000
#使用时序模型构建模型
model = keras.Sequential()
#添加嵌入层
model.add(layers.Embedding(vocab_size, 16))
```

```
#添加池化层,全局池化
model.add(layers.GlobalAveragePooling1D())
#添加密集连接层
model.add(layers.Dense(16, activation='relu'))
model.add(layers.Dense(1, activation='sigmoid'))
#输出各层的详细信息
model.summary()
#初始化模型参数
model.compile(optimizer='adam', loss='binary_crossentropy', metrics=['accuracy'])
```

(8) 代码的运行结果如下。

```
Model: " sequential "
_____
Layer (type)                 Output Shape              Param #
=================================================================
embedding (Embedding)        (None, None, 16)          160000
_____
global_average_pooling1d (Gl (None, 16)                0
_____
dense (Dense)                (None, 16)                272
_____
dense_1 (Dense)              (None, 1)                 17
=================================================================
Total params: 160,289
Trainable params: 160,289
Non-trainable params: 0
```

(9) 对成功构建的模型进行训练,并对结果进行验证,本例中的训练次数为40次,代码如下。

```
mport ssl
#引入 TensorFlow 模块并基于 TensorFlow 构建 Keras
import tensorflow as tf
import tensorflow.keras as keras
import tensorflow.keras.layers as layers
ssl._create_default_https_context = ssl._create_unverified_context
#导入数据集并划分测试集和训练集
imdb=keras.datasets.imdb
(train_x, train_y), (test_x, text_y)=keras.datasets.imdb.load_data(num_words=10000)
#将数据转换成字典
word_index = imdb.get_word_index()
word2id = {k:(v+3) for k, v in word_index.items()}
word2id['<PAD>'] = 0
word2id['<START>'] = 1
word2id['<UNK>'] = 2
word2id['<UNUSED>'] = 3
```

```python
id2word = {v:k for k, v in word2id.items()}
def get_words(sent_ids):
    return ' '.join([id2word.get(i, '?') for i in sent_ids])
sent = get_words(train_x[0])
#句子末尾padding
train_x = keras.preprocessing.sequence.pad_sequences(
    train_x, value=word2id['<PAD>'],
    padding='post', maxlen=256
)
test_x = keras.preprocessing.sequence.pad_sequences(
    test_x, value=word2id['<PAD>'],
    padding='post', maxlen=256
)
#定义词汇size
vocab_size = 10000
#使用时序模型构建模型
model = keras.Sequential()
#添加嵌入层
model.add(layers.Embedding(vocab_size, 16))
#添加池化层，全局池化
model.add(layers.GlobalAveragePooling1D())
#添加密集连接层
model.add(layers.Dense(16, activation='relu'))
model.add(layers.Dense(1, activation='sigmoid'))
#输出各层的详细信息
model.summary()
#初始化模型参数
model.compile(optimizer='adam', loss='binary_crossentropy', metrics=['accuracy'])
#划分训练数据
x_val = train_x[:10000]
x_train = train_x[10000:]
y_val = train_y[:10000]
y_train = train_y[10000:]
#制订训练计划并执行
history = model.fit(x_train,y_train, epochs=40, batch_size=512,
    validation_data=(x_val, y_val), verbose=1)
#对模型进行评估
result = model.evaluate(test_x, text_y)
#输出评估结果
print(result)
```

（10）代码的运行结果如下。

```
Model: "sequential"
_____
```

```
Layer (type)                 Output Shape              Param #
=================================================================
embedding (Embedding)        (None, None, 16)          160000
_____
global_average_pooling1d (Gl (None, 16)                0
_____
dense (Dense)                (None, 16)                272
_____
dense_1 (Dense)              (None, 1)                 17
=================================================================
Total params: 160,289
Trainable params: 160,289
Non-trainable params: 0
_____
Train on 15000 samples, validate on 10000 samples
Epoch 1/40
15000/15000 [==============================] - 2s 115us/sample - loss: 0.6920 - accuracy: 0.5419 - val_loss: 0.6903 - val_accuracy: 0.5644
...
Epoch 40/40
15000/15000 [==============================] - 1s 36us/sample - loss: 0.0949 - accuracy: 0.9741 - val_loss: 0.3097 - val_accuracy: 0.8835
25000/1 [====]
- 1s 47us/sample - loss: 0.3341 - accuracy: 0.8718
[0.33117988712787627, 0.8718]
```

（11）显示预测结果的损失率，代码如下。

```python
import ssl
#引入 TensorFlow 模块并基于 TensorFlow 构建 Keras
import tensorflow as tf
import tensorflow.keras as keras
import tensorflow.keras.layers as layers
ssl._create_default_https_context = ssl._create_unverified_context
#导入数据集并划分测试集和训练集
imdb=keras.datasets.imdb
(train_x, train_y), (test_x, text_y)=keras.datasets.imdb.load_data(num_words=10000)
#将数据转换成字典
word_index = imdb.get_word_index()
word2id = {k:(v+3) for k, v in word_index.items()}
word2id['<PAD>'] = 0
word2id['<START>'] = 1
word2id['<UNK>'] = 2
word2id['<UNUSED>'] = 3
id2word = {v:k for k, v in word2id.items()}
```

```python
def get_words(sent_ids):
    return ' '.join([id2word.get(i, '?') for i in sent_ids])
sent = get_words(train_x[0])
#句子末尾padding
train_x = keras.preprocessing.sequence.pad_sequences(
    train_x, value=word2id['<PAD>'],
    padding='post', maxlen=256
)
test_x = keras.preprocessing.sequence.pad_sequences(
    test_x, value=word2id['<PAD>'],
    padding='post', maxlen=256
)
#定义词汇size
vocab_size = 10000
#使用时序模型构建模型
model = keras.Sequential()
#添加嵌入层
model.add(layers.Embedding(vocab_size, 16))
#添加池化层，全局池化
model.add(layers.GlobalAveragePooling1D())
#添加密集连接层
model.add(layers.Dense(16, activation='relu'))
model.add(layers.Dense(1, activation='sigmoid'))
#输出各层的详细信息
model.summary()
#初始化模型参数
model.compile(optimizer='adam', loss='binary_crossentropy', metrics=['accuracy'])
#划分训练数据
x_val = train_x[:10000]
x_train = train_x[10000:]
y_val = train_y[:10000]
y_train = train_y[10000:]
#制订训练计划并执行
history = model.fit(x_train,y_train, epochs=40, batch_size=512,
    validation_data=(x_val, y_val), verbose=1)
#对模型进行评估
result = model.evaluate(test_x, text_y)
#导入图形化工具
import matplotlib.pyplot as plt
history_dict = history.history
history_dict.keys()
acc = history_dict['accuracy']
```

```
val_acc = history_dict['val_accuracy']
loss = history_dict['loss']
val_loss = history_dict['val_loss']
epochs = range(1, len(acc)+1)
#显示图像
plt.plot(epochs, loss, 'bo', label='training')
plt.plot(epochs, val_loss, 'b', label='validation')
plt.title('Training and validation loss')
plt.xlabel('epochs')
plt.ylabel('loss')
plt.legend()
plt.show()
```

（12）运行代码，得到预测结果的损失率如图 8-1 所示。

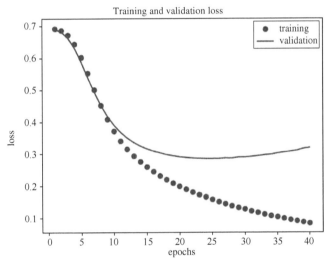

图 8-1　预测结果的损失率

（13）显示模型训练的精确度，代码如下。

```
import ssl
#引入 TensorFlow 模块并基于 TensorFlow 构建 Keras
import tensorflow as tf
import tensorflow.keras as keras
import tensorflow.keras.layers as layers
ssl._create_default_https_context = ssl._create_unverified_context
#导入数据集并划分训练集和测试集
imdb=keras.datasets.imdb
(train_x, train_y), (test_x, text_y)=keras.datasets.imdb.load_data(num_words=10000)
#将数据转换成字典
word_index = imdb.get_word_index()
word2id = {k:(v+3) for k, v in word_index.items()}
```

```python
word2id['<PAD>'] = 0
word2id['<START>'] = 1
word2id['<UNK>'] = 2
word2id['<UNUSED>'] = 3
id2word = {v:k for k, v in word2id.items()}
def get_words(sent_ids):
    return ' '.join([id2word.get(i, '?') for i in sent_ids])
sent = get_words(train_x[0])
#句子末尾padding
train_x = keras.preprocessing.sequence.pad_sequences(
    train_x, value=word2id['<PAD>'],
    padding='post', maxlen=256
)
test_x = keras.preprocessing.sequence.pad_sequences(
    test_x, value=word2id['<PAD>'],
    padding='post', maxlen=256
)
#定义词汇size
vocab_size = 10000
#使用时序模型构建模型
model = keras.Sequential()
#添加嵌入层
model.add(layers.Embedding(vocab_size, 16))
#添加池化层，全局池化
model.add(layers.GlobalAveragePooling1D())
#添加密集连接层
model.add(layers.Dense(16, activation='relu'))
model.add(layers.Dense(1, activation='sigmoid'))
#输出各层的详细信息
model.summary()
#初始化模型参数
model.compile(optimizer='adam', loss='binary_crossentropy', metrics=['accuracy'])
#划分训练数据
x_val = train_x[:10000]
x_train = train_x[10000:]
y_val = train_y[:10000]
y_train = train_y[10000:]
#制订训练计划并执行
history = model.fit(x_train,y_train, epochs=40, batch_size=512,
    validation_data=(x_val, y_val), verbose=1)
#对模型进行评估
result = model.evaluate(test_x, text_y)
#导入图形化工具
import matplotlib.pyplot as plt
```

```
history_dict = history.history
history_dict.keys()
acc = history_dict['accuracy']
val_acc = history_dict['val_accuracy']
loss = history_dict['loss']
val_loss = history_dict['val_loss']
epochs = range(1, len(acc)+1)
#清除所有轴,但是打开窗口,这样它可以被重复使用
plt.clf()
#显示图像
plt.plot(epochs, acc, 'bo', label='training')
plt.plot(epochs, val_acc, 'b', label='validation')
plt.title('Training and validation accuracy')
plt.xlabel('epochs')
plt.ylabel('accuracy')
plt.legend()
plt.show()
```

(14)运行代码,得到模型训练的精确度如图 8-2 所示。

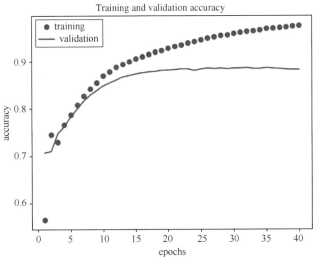

图 8-2 模型训练的精确度

8.2 卷积文本分类

本节探讨如何使用卷积神经网络对文本进行分类,这种方式在实际过程中大量应用于处理情感分类等任务。

(1)导入数据并测试,代码如下。

```
#导入 NumPy 模块
import numpy as np
```

```
#导入图形化工具
import matplotlib.pyplot as plt
#引入TensorFlow模块并基于TensorFlow构建Keras
from tensorflow import keras
from tensorflow.keras import layers
from tensorflow.keras.preprocessing.sequence import pad_sequences
#定义超参
num_features = 3000
sequence_length = 300
embedding_dimension = 100
#导入数据集并划分训练集和测试集
(x_train, y_train), (x_test, y_test) = keras.datasets.imdb.load_data(num_words=num_features)
print(x_train.shape)
print(x_test.shape)
print(y_train.shape)
print(y_test.shape)
```

（2）代码的运行结果如下。

```
(25000,)
(25000,)
(25000,)
(25000,)
```

（3）在数据结果显示正确的前提下，对数据进行序列预处理，代码如下。

```
#导入NumPy模块
import numpy as np
#导入图形化工具
import matplotlib.pyplot as plt
#引入TensorFlow模块并基于TensorFlow构建Keras
from tensorflow import keras
from tensorflow.keras import layers
from tensorflow.keras.preprocessing.sequence import pad_sequences
#定义超参
num_features = 3000
sequence_length = 300
embedding_dimension = 100
#导入数据集并划分训练集和测试集
(x_train, y_train), (x_test, y_test) = keras.datasets.imdb.load_data(num_words=num_features)
#序列化数据
x_train = pad_sequences(x_train, maxlen=sequence_length)
x_test = pad_sequences(x_test, maxlen=sequence_length)
print(x_train.shape)
print(x_test.shape)
```

第8章 TensorFlow 文本分类

```
print(y_train.shape)
print(y_test.shape)
```

说明：使用 pad_sequences 填充序列化数据，并形成一个长度相同的新序列。

（4）代码的运行结果如下。

```
(25000, 300)
(25000, 300)
(25000,)
(25000,)
```

（5）构造基本句子分类器，代码如下。

```
#导入 NumPy 模块
import numpy as np
#导入图形化工具
import matplotlib.pyplot as plt
#引入 TensorFlow 模块并基于 TensorFlow 构建 Keras
from tensorflow import keras
from tensorflow.keras import layers
from tensorflow.keras.preprocessing.sequence import pad_sequences
#定义超参
num_features = 3000
sequence_length = 300
embedding_dimension = 100
#导入数据集并划分训练集和测试集
(x_train, y_train), (x_test, y_test) = keras.datasets.imdb.load_data(num_words=num_features)
#序列化数据
x_train = pad_sequences(x_train, maxlen=sequence_length)
x_test = pad_sequences(x_test, maxlen=sequence_length)
#定义 CNN 模型函数
def imdb_cnn():
#使用时序模型构建模型
    model = keras.Sequential([
#嵌入层
        layers.Embedding(input_dim=num_features,
            output_dim=embedding_dimension,input_length=sequence_length),
#池化层
        layers.Conv1D(filters=50, kernel_size=5, strides=1, padding='valid'),
#在 steps 维度（也就是第二维）求最大值，但是限制每一步的池化大小
        layers.MaxPool1D(2, padding='valid'),
#在全连接之前进行处理
        layers.Flatten(),
#全连接层
        layers.Dense(10, activation='relu'),
        layers.Dense(1, activation='sigmoid')
```

```
    ])
    #初始化模型参数
    model.compile(optimizer=keras.optimizers.Adam(1e-3),
        loss=keras.losses.BinaryCrossentropy(), metrics=['accuracy'])
    return model
#使用模型函数构建模型
model = imdb_cnn()
#输出模型详情
model.summary()
```

(6) 代码的运行结果如下。

```
Model: "sequential"
_____
Layer (type)                 Output Shape              Param #
=================================================================
embedding (Embedding)        (None, 300, 100)          300000
_____
conv1d (Conv1D)              (None, 296, 50)           25050
_____
max_pooling1d (MaxPooling1D) (None, 148, 50)           0
_____
flatten (Flatten)            (None, 7400)              0
_____
dense (Dense)                (None, 10)                74010
_____
dense_1 (Dense)              (None, 1)                 11
=================================================================
Total params: 399,071
Trainable params: 399,071
Non-trainable params: 0
_____
```

(7) 对模型进行训练，本例中训练 5 次，代码如下。

```
#导入 NumPy 模块
import numpy as np
#导入图形化工具
import matplotlib.pyplot as plt
#引入 TensorFlow 模块并基于 TensorFlow 构建 Keras
from tensorflow import keras
from tensorflow.keras import layers
from tensorflow.keras.preprocessing.sequence import pad_sequences
#定义超参
num_features = 3000
sequence_length = 300
embedding_dimension = 100
```

```python
#导入数据集并划分训练集和测试集
(x_train, y_train), (x_test, y_test) = keras.datasets.imdb.load_data(num_words=num_features)
#序列化数据
x_train = pad_sequences(x_train, maxlen=sequence_length)
x_test = pad_sequences(x_test, maxlen=sequence_length)
#定义 CNN 模型函数
def imdb_cnn():
#使用时序模型构建模型
    model = keras.Sequential([
#嵌入层
        layers.Embedding(input_dim=num_features,
            output_dim=embedding_dimension,input_length=sequence_length),
#池化层
        layers.Conv1D(filters=50, kernel_size=5, strides=1, padding='valid'),
#在 steps 维度(也就是第二维)求最大值,但是限制每一步的池化大小。
        layers.MaxPool1D(2, padding='valid'),
#在全连接之前进行处理
        layers.Flatten(),
#全连接层
        layers.Dense(10, activation='relu'),
        layers.Dense(1, activation='sigmoid')
    ])
#初始化模型参数
    model.compile(optimizer=keras.optimizers.Adam(1e-3),
        loss=keras.losses.BinaryCrossentropy(), metrics=['accuracy'])
    return model
#使用模型函数构建模型
model = imdb_cnn()
#制订训练计划并执行
history = model.fit(x_train, y_train, batch_size=64, epochs=5, validation_split=0.1)
```

(8)代码的运行结果如下。

```
Train on 22500 samples, validate on 2500 samples
Epoch 1/5
22500/22500 [==============================] - 32s 1ms/sample - loss: 0.4527 - accuracy: 0.7557 - val_loss: 0.3020 - val_accuracy: 0.8724
...
Epoch 5/5
22500/22500 [==============================] - 32s 1ms/sample - loss: 0.0723 - accuracy: 0.9805 - val_loss: 0.3947 - val_accuracy: 0.8664
```

(9)显示图像,代码如下。

```
#导入 NumPy 模块
import numpy as np
```

```python
#导入图形化工具
import matplotlib.pyplot as plt
#引入TensorFlow模块并基于TensorFlow构建Keras
from tensorflow import keras
from tensorflow.keras import layers
from tensorflow.keras.preprocessing.sequence import pad_sequences
#定义超参
num_features = 3000
sequence_length = 300
embedding_dimension = 100
#导入数据集并划分训练集和测试集
(x_train, y_train), (x_test, y_test) = keras.datasets.imdb.load_data(num_words=num_features)
#序列化数据
x_train = pad_sequences(x_train, maxlen=sequence_length)
x_test = pad_sequences(x_test, maxlen=sequence_length)
#定义CNN模型函数
def imdb_cnn():
#使用时序模型构建模型
    model = keras.Sequential([
#嵌入层
        layers.Embedding(input_dim=num_features,
            output_dim=embedding_dimension,input_length=sequence_length),
#池化层
        layers.Conv1D(filters=50, kernel_size=5, strides=1, padding='valid'),
#在steps维度(也就是第二维)求最大值,但是限制每一步的池化大小
        layers.MaxPool1D(2, padding='valid'),
#在全连接之前进行处理
        layers.Flatten(),
#全连接层
        layers.Dense(10, activation='relu'),
        layers.Dense(1, activation='sigmoid')
    ])
#初始化模型参数
    model.compile(optimizer=keras.optimizers.Adam(1e-3),
        loss=keras.losses.BinaryCrossentropy(), metrics=['accuracy'])
    return model
#使用模型函数构建模型
model = imdb_cnn()
#制订训练计划并执行
history = model.fit(x_train, y_train, batch_size=64, epochs=5, validation_split=0.1)
#显示图像
plt.plot(history.history['accuracy'])
plt.plot(history.history['val_accuracy'])
```

```
plt.legend(['training', 'valiation'], loc='upper left')
plt.show()
```

（10）运行代码，得到卷积文本分类的训练结果如图8-3所示。

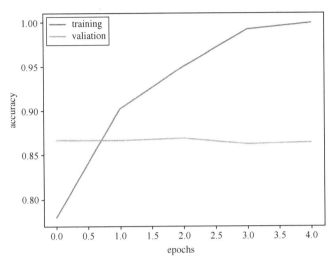

图8-3　卷积文本分类的训练结果

（11）在得到结果后，使用多核卷积网络对文本进行分类，代码如下。

```
#导入NumPy模块
import numpy as np
#导入图形化工具
import matplotlib.pyplot as plt
#引入TensorFlow模块并基于TensorFlow构建Keras
from tensorflow import keras
from tensorflow.keras import layers
from tensorflow.keras.preprocessing.sequence import pad_sequences
#定义超参
num_features = 3000
sequence_length = 300
embedding_dimension = 100
#导入数据集并划分训练集和测试集
(x_train, y_train), (x_test, y_test) = keras.datasets.imdb.load_data(num_words=num_features)
#序列化数据
x_train = pad_sequences(x_train, maxlen=sequence_length)
x_test = pad_sequences(x_test, maxlen=sequence_length)
#定义输出通道数
filter_sizes=[3,4,5]
#定义卷积模型函数
def convolution():
    inn = layers.Input(shape=(sequence_length, embedding_dimension, 1))
```

```
            cnns = []
            for size in filter_sizes:
                conv = layers.Conv2D(filters=64, kernel_size=(size, embedding_dimension),
                            strides=1, padding='valid', activation='relu')(inn)
                pool = layers.MaxPool2D(pool_size=(sequence_length-size+1, 1), padding='valid')(conv)
                cnns.append(pool)
            outt = layers.concatenate(cnns)
        #构建模型
            model = keras.Model(inputs=inn, outputs=outt)
            return model
        #定义多核卷积网络函数
        def cnn_mulfilter():
        #使用时序模型构建模型
            model = keras.Sequential([
        #嵌入层
                layers.Embedding(input_dim=num_features, output_dim=embedding_dimension,
                    input_length=sequence_length),
        #将输入调整为特定的尺寸
                layers.Reshape((sequence_length, embedding_dimension, 1)),
        #调用卷积模型函数
                convolution(),
        #在全连接之前进行处理
                layers.Flatten(),
        #全连接层
                layers.Dense(10, activation='relu'),
                layers.Dropout(0.2),
                layers.Dense(1, activation='sigmoid')
            ])
        #初始化模型参数
            model.compile(optimizer=keras.optimizers.Adam(),
                loss=keras.losses.BinaryCrossentropy(), metrics=['accuracy'])
            return model
        #使用多核卷积网络函数构建模型
        model = cnn_mulfilter()
        #输出模型详情
        model.summary()
```

（12）代码的运行结果如下。

```
Model: "sequential"
_____
Layer (type)                 Output Shape              Param #
=================================================================
embedding (Embedding)        (None, 300, 100)          300000
```

```
_____
reshape (Reshape)              (None, 300, 100, 1)         0
_____
model (Model)                  (None, 1, 1, 192)           76992
_____
flatten (Flatten)              (None, 192)                 0
_____
dense (Dense)                  (None, 10)                  1930
_____
dropout (Dropout)              (None, 10)                  0
_____
dense_1 (Dense)                (None, 1)                   11
================================================================
Total params: 378,933
Trainable params: 378,933
Non-trainable params: 0
_____
```

（13）对模型进行训练，本例中训练5次，代码如下。

```
#导入NumPy模块
import numpy as np
#导入图形化工具
import matplotlib.pyplot as plt
#引入TensorFlow模块并基于TensorFlow构建Keras
from tensorflow import keras
from tensorflow.keras import layers
from tensorflow.keras.preprocessing.sequence import pad_sequences
#定义超参
num_features = 3000
sequence_length = 300
embedding_dimension = 100
#导入数据集并划分训练集和测试集
(x_train, y_train), (x_test, y_test) = keras.datasets.imdb.load_data(num_words=num_features)
#序列化数据
x_train = pad_sequences(x_train, maxlen=sequence_length)
x_test = pad_sequences(x_test, maxlen=sequence_length)
#定义输出通道数
filter_sizes=[3,4,5]
#定义卷积模型函数
def convolution():
    inn = layers.Input(shape=(sequence_length, embedding_dimension, 1))
    cnns = []
    for size in filter_sizes:
```

```
                conv = layers.Conv2D(filters=64, kernel_size=(size, embedding_dimension),
                            strides=1, padding='valid', activation='relu')(inn)
                pool = layers.MaxPool2D(pool_size=(sequence_length-size+1, 1), padding='valid')(conv)
                cnns.append(pool)
            outt = layers.concatenate(cnns)
#构建模型
            model = keras.Model(inputs=inn, outputs=outt)
            return model
#定义多核卷积网络函数
def cnn_mulfilter():
#使用时序模型构建模型
            model = keras.Sequential([
#嵌入层
                layers.Embedding(input_dim=num_features, output_dim=embedding_dimension,
                    input_length=sequence_length),
#将输入调整为特定的尺寸
                layers.Reshape((sequence_length, embedding_dimension, 1)),
#调用卷积模型函数
                convolution(),
#在全链接之前进行处理
                layers.Flatten(),
#全连接层
                layers.Dense(10, activation='relu'),
                layers.Dropout(0.2),
                layers.Dense(1, activation='sigmoid')
            ])
#初始化模型参数
            model.compile(optimizer=keras.optimizers.Adam(),
                loss=keras.losses.BinaryCrossentropy(), metrics=['accuracy'])
            return model
#使用多核卷积网络函数构建模型
model = cnn_mulfilter()
#制订训练计划并执行
history = model.fit(x_train, y_train, batch_size=64, epochs=5, validation_split=0.1)
```

（14）代码的运行结果如下。

```
Train on 22500 samples, validate on 2500 samples
Epoch 1/5
22500/22500 [==============================] - 115s 5ms/sample - loss: 0.4822 - accuracy: 0.7530 - val_loss: 0.3170 - val_accuracy: 0.8728
    ...
Epoch 5/5
22500/22500 [==============================] - 97s 4ms/sample - loss: 0.0911 - accuracy:
```

0.9753 - val_loss: 0.3345 - val_accuracy: 0.8876

（15）显示图像，代码如下。

```
#导入NumPy模块
import numpy as np
#导入图形化工具
import matplotlib.pyplot as plt
#引入TensorFlow模块并基于TensorFlow构建Keras
from tensorflow import keras
from tensorflow.keras import layers
from tensorflow.keras.preprocessing.sequence import pad_sequences
#定义超参
num_features = 3000
sequence_length = 300
embedding_dimension = 100
#导入数据集并划分训练集和测试集
(x_train, y_train), (x_test, y_test) = keras.datasets.imdb.load_data(num_words=num_features)
#序列化数据
x_train = pad_sequences(x_train, maxlen=sequence_length)
x_test = pad_sequences(x_test, maxlen=sequence_length)
#定义输出通道数
filter_sizes=[3,4,5]
#定义卷积模型函数
def convolution():
    inn = layers.Input(shape=(sequence_length, embedding_dimension, 1))
    cnns = []
    for size in filter_sizes:
        conv = layers.Conv2D(filters=64, kernel_size=(size, embedding_dimension),
                        strides=1, padding='valid', activation='relu')(inn)
        pool = layers.MaxPool2D(pool_size=(sequence_length-size+1, 1), padding='valid')(conv)
        cnns.append(pool)
    outt = layers.concatenate(cnns)
#构建模型
    model = keras.Model(inputs=inn, outputs=outt)
    return model
#定义多核卷积网络函数
def cnn_mulfilter():
#使用时序模型构建模型
    model = keras.Sequential([
#嵌入层
        layers.Embedding(input_dim=num_features, output_dim=embedding_dimension,
            input_length=sequence_length),
```

```
        #将输入调整为特定的尺寸
        layers.Reshape((sequence_length, embedding_dimension, 1)),
        #调用卷积模型函数
        convolution(),
        #在全连接之前进行处理
        layers.Flatten(),
        #全连接层
        layers.Dense(10, activation='relu'),
        layers.Dropout(0.2),
        layers.Dense(1, activation='sigmoid')
    ])
    #初始化模型参数
    model.compile(optimizer=keras.optimizers.Adam(),
        loss=keras.losses.BinaryCrossentropy(), metrics=['accuracy'])
    return model
#使用多核卷积网络函数构建模型
model = cnn_mulfilter()
#制订训练计划并执行
history = model.fit(x_train, y_train, batch_size=64, epochs=5, validation_split=0.1)
#显示图像
plt.plot(history.history['accuracy'])
plt.plot(history.history['val_accuracy'])
plt.legend(['training', 'valiation'], loc='upper left')
plt.show()
```

（16）运行代码，得到使用多核卷积网络进行文本分类的训练结果如图 8-4 所示。

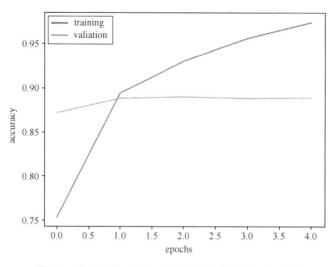

图 8-4　使用多核卷积网络进行文本分类的训练结果

可以看出，使用多核卷积网络进行文本分类的训练结果有明显变化。在调优的过程中可以考虑用这种方法提高精确度。

8.3 RNN 文本分类

循环神经网络（RNN）模型非常流行，在自然语言处理（NLP）的很多任务中得到了应用。基于 RNN 的语言模型主要有两个方面的应用。

第一，根据实际情况评估语法和语义的精确度，大量应用于机器翻译系统中。

第二，构建类似风格的语言模型，生成对应风格的新语言，比较典型的应用是根据著名模型的风格生成新的模型。

本节使用 RNN 进行文本分类，步骤如下。

（1）使用 tensorflow_datasets 构造输入数据集，并划分训练集和测试集，在获取 tokenizer 对象后，进行字符处理及 ID 转换，代码如下。

```
#导入数据集
import tensorflow_datasets as tfds
dataset, info = tfds.load('imdb_reviews/subwords8k', with_info=True,
    as_supervised=True)
train_dataset, test_dataset = dataset['train'], dataset['test']
#对 tokenizer 对象进行字符处理及 ID 转换
tokenizer = info.features['text'].encoder
#输出处理后的对象大小
print('vocabulary size: ', tokenizer.vocab_size)
```

（2）代码的运行结果如下。

```
vocabulary size:  8185
```

（3）对 tokenizer 对象进行测试，代码如下。

```
#导入数据集
import tensorflow_datasets as tfds
dataset, info = tfds.load('imdb_reviews/subwords8k', with_info=True,
    as_supervised=True)
train_dataset, test_dataset = dataset['train'], dataset['test']
#对 tokenizer 对象进行字符处理及 ID 转换
tokenizer = info.features['text'].encoder
#输出处理后的对象大小
print('vocabulary size: ', tokenizer.vocab_size)
#定义测试数据
sample_string = 'Hello word , TensorFlow2.0'
#使用测试数据进行转换测试
tokenized_string = tokenizer.encode(sample_string)
print('tokened id: ', tokenized_string)
#解码还原字符串
src_string = tokenizer.decode(tokenized_string)
print('original string: ', src_string)
```

(4) 代码的运行结果如下。

```
vocabulary size: 8185
tokened id: [4025, 222, 2621, 1199, 6307, 2327, 2934, 7979, 7975, 7977]
original string: Hello word , TensorFlow2.0
```

(5) 解析出每个对象,代码如下。

```
#导入数据集
import tensorflow_datasets as tfds
dataset, info = tfds.load('imdb_reviews/subwords8k', with_info=True,
    as_supervised=True)
train_dataset, test_dataset = dataset['train'], dataset['test']
#对tokenizer对象进行字符处理及ID转换
tokenizer = info.features['text'].encoder
#输出处理后的对象大小
print('vocabulary size: ', tokenizer.vocab_size)
#定义测试数据
sample_string = 'Hello word , TensorFlow2.0'
#使用测试数据进行转换测试
tokenized_string = tokenizer.encode(sample_string)
print('tokened id: ', tokenized_string)
#解码还原字符串
src_string = tokenizer.decode(tokenized_string)
print('original string: ', src_string)
#解码字符串
src_string = tokenizer.decode(tokenized_string)
print('original string: ', src_string)
for t in tokenized_string:
    print(str(t)+'->['+tokenizer.decode([t])+ ']')
```

(6) 代码的运行结果如下。

```
vocabulary size: 8185
tokened id: [4025, 222, 2621, 1199, 6307, 2327, 2934, 7979, 7975, 7977]
original string: Hello word , TensorFlow2.0
4025->[Hell]
222->[o ]
2621->[word]
1199->[ , ]
6307->[Ten]
2327->[sor]
2934->[flow]
7979->[2]
7975->[.]
7977->[0]
```

(7) 在数据正确的前提下,构建模型并对其进行训练,本例中训练2次,代码如下。

```python
#导入数据集
import tensorflow_datasets as tfds
dataset, info = tfds.load('imdb_reviews/subwords8k', with_info=True,
    as_supervised=True)
train_dataset, test_dataset = dataset['train'], dataset['test']
#对 tokenizer 对象进行字符处理及 ID 转换
tokenizer = info.features['text'].encoder
#输出处理后的对象大小
print('vocabulary size: ', tokenizer.vocab_size)
#定义测试数据
sample_string = 'Hello word , TensorFlow2.0'
#使用测试数据进行转换测试
tokenized_string = tokenizer.encode(sample_string)
print('tokened id: ', tokenized_string)
#解码还原字符串
src_string = tokenizer.decode(tokenized_string)
print('original string: ', src_string)
#解码字符串
src_string = tokenizer.decode(tokenized_string)
#构建训练常数
BUFFER_SIZE=10000
BATCH_SIZE = 64
#按照定义的常数打乱数据
train_dataset = train_dataset.shuffle(BUFFER_SIZE)
#根据 BATCH_SIZE 合成 batch
train_dataset = train_dataset.padded_batch(BATCH_SIZE, train_dataset.output_shapes)
test_dataset = test_dataset.padded_batch(BATCH_SIZE, test_dataset.output_shapes)
#定义模型函数
def get_model():
#使用时序模型构建模型
    model = tf.keras.Sequential([
#嵌入层
        tf.keras.layers.Embedding(tokenizer.vocab_size, 64),
#RNN 的双向封装器,对序列进行前向和后向计算
        tf.keras.layers.Bidirectional(tf.keras.layers.LSTM(64)),
#全连接层
        tf.keras.layers.Dense(64, activation='relu'),
        tf.keras.layers.Dense(1, activation='sigmoid')
    ])
    return model
#使用定义模型构建模型
model = get_model()
#初始化模型函数
model.compile(loss='binary_crossentropy', optimizer='adam', metrics=['accuracy'])
```

```python
#执行模型训练计划并训练2次
history = model.fit(train_dataset, epochs=2,
                validation_data=test_dataset)
```

说明：因为此处的句子长度会逐渐增加，所以只能使用序列模型，不能使用 Keras 的函数 API。

（8）代码的运行结果如下。

```
vocabulary size:  8185
Epoch 1/2
91/391 [==============================] - 1021s 3s/step - loss: 0.5475 - accuracy: 0.7183 - val_loss: 0.0000e+00 - val_accuracy: 0.0000e+00
Epoch 2/2
390/391 [============================>.] - ETA: 2s - loss: 0.3836 - accuracy: 0.8450
```

（9）查看训练过程，显示精确度，代码如下。

```python
#导入数据集
import tensorflow_datasets as tfds
dataset, info = tfds.load('imdb_reviews/subwords8k', with_info=True,
        as_supervised=True)
train_dataset, test_dataset = dataset['train'], dataset['test']
#对 tokenizer 对象进行字符处理及 ID 转换
tokenizer = info.features['text'].encoder
#输出处理后的对象大小
print('vocabulary size: ', tokenizer.vocab_size)
#定义测试数据
sample_string = 'Hello word , TensorFlow2.0'
#使用测试数据进行转换测试
tokenized_string = tokenizer.encode(sample_string)
print('tokened id: ', tokenized_string)
#解码还原字符串
src_string = tokenizer.decode(tokenized_string)
print('original string: ', src_string)
#解码字符串
src_string = tokenizer.decode(tokenized_string)
#构建训练常数
BUFFER_SIZE=10000
BATCH_SIZE = 64
#按照定义的常数打乱数据
train_dataset = train_dataset.shuffle(BUFFER_SIZE)
#根据 BATCH_SIZE 合成 batch
train_dataset = train_dataset.padded_batch(BATCH_SIZE, train_dataset.output_shapes)
test_dataset = test_dataset.padded_batch(BATCH_SIZE, test_dataset.output_shapes)
#定义模型函数
def get_model():
#使用时序模型构建模型
    model = tf.keras.Sequential([
```

```python
#嵌入层
        tf.keras.layers.Embedding(tokenizer.vocab_size, 64),
#RNN 的双向封装器,对序列进行前向和后向计算
        tf.keras.layers.Bidirectional(tf.keras.layers.LSTM(64)),
#全连接层
        tf.keras.layers.Dense(64, activation='relu'),
        tf.keras.layers.Dense(1, activation='sigmoid')
    ])
    return model
#使用定义模型构建模型
model = get_model()
#初始化模型函数
model.compile(loss='binary_crossentropy', optimizer='adam', metrics=['accuracy'])
#执行模型训练计划并训练2次
history = model.fit(train_dataset, epochs=2,
                    validation_data=test_dataset)
#导入图形化工具
import matplotlib.pyplot as plt
#显示图像
def plot_graphs(history, string):
    plt.plot(history.history[string])
    plt.plot(history.history['val_'+string])
    plt.xlabel('epochs')
    plt.ylabel(string)
    plt.legend([string, 'val_'+string])
    plt.show()
#显示训练过程与精确度的关系
plot_graphs(history, 'accuracy')
```

（10）运行代码，得到训练过程与精确度的关系如图 8-5 所示。

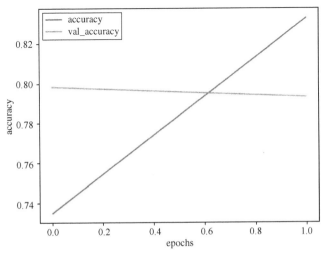

图 8-5　训练过程与精确度的关系

(11）查看训练过程，显示损失率，代码如下。

```
#导入数据集
import tensorflow_datasets as tfds
dataset, info = tfds.load('imdb_reviews/subwords8k', with_info=True,
    as_supervised=True)
train_dataset, test_dataset = dataset['train'], dataset['test']
#对 tokenizer 对象进行字符处理及 ID 转换
tokenizer = info.features['text'].encoder
#输出处理后的对象大小
print('vocabulary size: ', tokenizer.vocab_size)
#定义测试数据
sample_string = 'Hello word , TensorFlow2.0'
#使用测试数据进行转换测试
tokenized_string = tokenizer.encode(sample_string)
print('tokened id: ', tokenized_string)
#解码还原字符串
src_string = tokenizer.decode(tokenized_string)
print('original string: ', src_string)
#解码字符串
src_string = tokenizer.decode(tokenized_string)
#构建训练常数
BUFFER_SIZE=10000
BATCH_SIZE = 64
#按照定义的常数打乱数据
train_dataset = train_dataset.shuffle(BUFFER_SIZE)
#根据 BATCH_SIZE 合成 batch
train_dataset = train_dataset.padded_batch(BATCH_SIZE, train_dataset.output_shapes)
test_dataset = test_dataset.padded_batch(BATCH_SIZE, test_dataset.output_shapes)
#定义模型函数
def get_model():
#使用时序模型构建模型
    model = tf.keras.Sequential([
#嵌入层
        tf.keras.layers.Embedding(tokenizer.vocab_size, 64),
#RNN 的双向封装器,对序列进行前向和后向计算
        tf.keras.layers.Bidirectional(tf.keras.layers.LSTM(64)),
#全连接层
        tf.keras.layers.Dense(64, activation='relu'),
        tf.keras.layers.Dense(1, activation='sigmoid')
    ])
    return model
#使用定义模型构建模型
model = get_model()
```

```
#初始化模型函数
model.compile(loss='binary_crossentropy', optimizer='adam', metrics=['accuracy'])
#执行模型训练计划并训练2次
history = model.fit(train_dataset, epochs=2, validation_data=test_dataset)
#导入图形化工具
import matplotlib.pyplot as plt
#显示图像
def plot_graphs(history, string):
    plt.plot(history.history[string])
    plt.plot(history.history['val_'+string])
    plt.xlabel('epochs')
    plt.ylabel(string)
    plt.legend([string, 'val_'+string])
    plt.show()
#显示训练过程与损失率的关系
plot_graphs(history, 'loss')
```

(12)运行代码,得到训练过程与损失率的关系如图8-6所示。

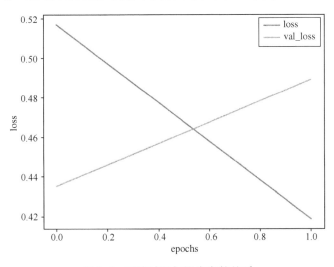

图 8-6　训练过程与损失率的关系

(13)因为前面构建的模型没有 mask 掉序列的 padding,所以与没有 padding 的情况相比,在有 padding 的情况下进行训练可能会出现偏差。下面对这两种情况进行对比,代码如下。

```
#导入数据集
import tensorflow_datasets as tfds
dataset, info = tfds.load('imdb_reviews/subwords8k', with_info=True,
    as_supervised=True)
train_dataset, test_dataset = dataset['train'], dataset['test']
#对 tokenizer 对象进行字符处理及 ID 转换
tokenizer = info.features['text'].encoder
```

```python
#输出处理后的对象大小
print('vocabulary size: ', tokenizer.vocab_size)
#定义测试数据
sample_string = 'Hello word , TensorFlow2.0'
#使用测试数据进行转换测试
tokenized_string = tokenizer.encode(sample_string)
print('tokened id: ', tokenized_string)
#解码还原字符串
src_string = tokenizer.decode(tokenized_string)
print('original string: ', src_string)
#解码字符串
src_string = tokenizer.decode(tokenized_string)
#构建训练常数
BUFFER_SIZE=10000
BATCH_SIZE = 64
#按照定义的常数打乱数据
train_dataset = train_dataset.shuffle(BUFFER_SIZE)
#根据 BATCH_SIZE 合成 batch
train_dataset = train_dataset.padded_batch(BATCH_SIZE, train_dataset.output_shapes)
test_dataset = test_dataset.padded_batch(BATCH_SIZE, test_dataset.output_shapes)
#定义模型函数
def get_model():
#使用时序模型构建模型
    model = tf.keras.Sequential([
#嵌入层
        tf.keras.layers.Embedding(tokenizer.vocab_size, 64),
#RNN 的双向封装器,对序列进行前向和后向计算
        tf.keras.layers.Bidirectional(tf.keras.layers.LSTM(64)),
#全连接层
        tf.keras.layers.Dense(64, activation='relu'),
        tf.keras.layers.Dense(1, activation='sigmoid')
    ])
    return model
#使用定义模型构建模型
model = get_model()
#初始化模型参数
model.compile(loss='binary_crossentropy', optimizer='adam', metrics=['accuracy'])
#执行训练计划并训练 2 次
history = model.fit(train_dataset, epochs=2, validation_data=test_dataset)
#定义填充函数
def pad_to_size(vec, size):
    zeros = [0] * (size-len(vec))
    vec.extend(zeros)
    return vec
```

```
#定义预测函数
def sample_predict(sentence, pad=False):
    tokened_sent = tokenizer.encode(sentence)
    if pad:
        tokened_sent = pad_to_size(tokened_sent, 64)
    pred = model.predict(tf.expand_dims(tokened_sent, 0))
    return pred
#没有padding的情况
sample_pred_text = ('The movie was cool. The animation and the graphics '
    'were out of this world. I would recommend this movie.')
predictions = sample_predict(sample_pred_text, pad=False)
print(predictions)
#有padding的情况
sample_pred_text = ('The movie was cool. The animation and the graphics '
    'were out of this world. I would recommend this movie.')
predictions = sample_predict(sample_pred_text, pad=True)
print (predictions)
```

（14）代码的运行结果如下。

```
[0.6481597]]
[0.6016696]]
```

比较有 padding 和没有 padding 的情况，可以看出预测的结果会受 padding 的影响。

本章通过几个例子对文本分类的机器学习过程进行了介绍，希望读者能够对本章的内容和前面学习过的内容进行总结。

文本分类是机器学习中最重要的部分之一，本章通过简单的例子说明了如何使用 TensorFlow 2.0 构建一个简单的文本分类系统。在实际使用中，应该根据实际情况构建模型和选择参数，以达到更好的效果。

第 9 章 TensorFlow 图像处理

图像处理是机器学习的重要应用，本书以 TensorFlow 2.0 为主，使用两个例子来介绍图像处理。

9.1 图像分类

本节使用 Fashion MNIST 数据集，该数据集包含 10 个类别中的 70000 个灰度图像。图像显示了低分辨率（28×28 像素）的单件服装，通常作为计算机视觉机器学习的基础数据集。

（1）导入 Fashion MNIST 数据集，代码如下。

```
import ssl
#以 TensorFlow 为基础构建 Keras
from tensorflow import keras
from tensorflow.keras import layers
#导入图形化工具
import matplotlib.pyplot as plt
ssl._create_default_https_context = ssl._create_unverified_context
#导入数据集并划分训练集和测试集
(train_images, train_labels), (test_images, test_labels) = keras.datasets.fashion_mnist.load_data()
#输出训练集和测试集的内容
print(train_labels,test_labels)
#输出第 1 个图像
plt.imshow(train_images[0].reshape(28, 28))
plt.show()
```

（2）代码的运行结果如下。

```
ownloading data from https://storage.googleapis.com/tensorflow/tf-keras-datasets/train-labels-idx1-ubyte.gz
32768/29515 [==============] - 0s 2us/step
Downloading data from https://storage.googleapis.com/tensorflow/tf-keras-datasets/train-images-idx3-ubyte.gz
26427392/26421880 [============] - 73s 3us/step
Downloading data from https://storage.googleapis.com/tensorflow/tf-keras-datasets/t10k-labels-idx1-ubyte.gz
8192/5148 [===========] - 0s 0us/step
Downloading data from https://storage.googleapis.com/tensorflow/tf-keras-datasets/
```

```
t10k-images-idx3-ubyte.gz
    4423680/4422102 [============] - 13s 3us/step
    [9 0 0 ... 3 0 5] [9 2 1 ... 8 1 5]
```

（3）运行代码，得到数据集中的示例图像如图 9-1 所示。

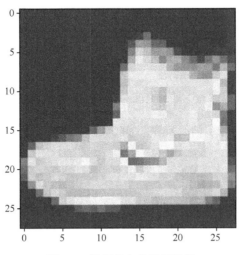

图 9-1　数据集中的示例图像

在 Fashion MNIST 数据集中，所有图像都是 28×28 NumPy 数组，像素为 0～255。其标签是一个整数数组，范围为 0～9。这些数组与图像所代表的服装类别相对应，类别标签对应表如表 9-1 所示。

表 9-1　类别标签对应表

数组值	服装类别	数组值	服装类别
0	T-shirt/top	5	Sandal
1	Trouser	6	Shirt
2	Pullover	7	Sneaker
3	Dress	8	Bag
4	Coat	9	Ankle boot

在 Fashion MNIST 数据集中，每个图像都映射到一类标签。由于类名不包含在数据集中，所以在使用过程中需要建立对应关系以便在绘制图像时使用，本例中进行如下定义。

```
class_names = ['T-shirt/top', 'Trouser', 'Pullover', 'Dress', 'Coat',
               'Sandal', 'Shirt', 'Sneaker', 'Bag', 'Ankle boot']
```

（4）在处理数据集之前，一般会对数据进行探索，在这个过程中主要关注数据格式。在本例中使用如下代码进行数据探索。

```
import ssl
#以 TensorFlow 为基础构建 Keras
from tensorflow import keras
from tensorflow.keras import layers
```

```
#导入图形化工具
import matplotlib.pyplot as plt
ssl._create_default_https_context = ssl._create_unverified_context
#导入数据集并划分训练集和测试集
(train_images, train_labels), (test_images, test_labels) = keras.datasets.fashion_mnist.load_data()
#导入数据集,包含70000张灰度图像,10个类别
print(train_images.shape)
print(train_labels.shape)
print(test_images.shape)
print(test_labels.shape)
```

(5)运行代码后,结果显示训练集中有60000个图像,每个图像的像素为28×28。

```
(60000, 28, 28)
(60000,)
(10000, 28, 28)
(10000,)
```

(6)在构建模型和进行训练之前,需要对数据集进行处理,以保证模型的准确性及训练的精准度。

```
import ssl
#以TensorFlow为基础构建Keras
from tensorflow import keras
from tensorflow.keras import layers
#导入图形化工具
import matplotlib.pyplot as plt
ssl._create_default_https_context = ssl._create_unverified_context
#导入数据集并划分训练集和测试集
(train_images, train_labels), (test_images, test_labels) = keras.datasets.fashion_mnist.load_data()
#显示全数据格式的图像
plt.figure()
plt.imshow(train_images[0])
plt.colorbar()
plt.grid(False)
plt.show()
```

(7)运行代码,得到全数据格式的示例图像如图9-2所示。

(8)将图像和分类对应并显示图像,代码如下。

```
import ssl
#以TensorFlow为基础构建Keras
from tensorflow import keras
from tensorflow.keras import layers
```

```
#导入图形化工具
import matplotlib.pyplot as plt
ssl._create_default_https_context = ssl._create_unverified_context
#导入数据集并划分训练集和测试集
(train_images, train_labels), (test_images, test_labels) = keras.datasets.fashion_mnist.load_data()
#定义分类信息
class_names = ['T-shirt/top', 'Trouser', 'Pullover', 'Dress', 'Coat',
               'Sandal', 'Shirt', 'Sneaker', 'Bag', 'Ankle boot']
#格式化训练图像
train_images = train_images / 255.0
test_images = test_images / 255.0
#将图像和分类进行对应并显示图像
plt.figure(figsize=(10,10))
for i in range(25):
    plt.subplot(5,5,i+1)
    plt.xticks([])
    plt.yticks([])
    plt.grid(False)
    plt.imshow(train_images[i], cmap=plt.cm.binary)
    plt.xlabel(class_names[train_labels[i]])
plt.show()
```

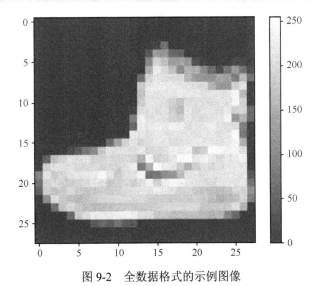

图 9-2 全数据格式的示例图像

（9）运行代码，得到图像和分类的对应图如图 9-3 所示。

图 9-3 图像和分类的对应图

（10）至此，完成了对 Fashion MNIST 数据集的简单处理，下面构造相应的网络对其进行分析，代码如下。

```
import ssl
#以 TensorFlow 为基础构建 Keras
from tensorflow import keras
from tensorflow.keras import layers
#导入图形化工具
import matplotlib.pyplot as plt
ssl._create_default_https_context = ssl._create_unverified_context
#导入数据集并划分训练集和测试集
(train_images, train_labels), (test_images, test_labels) = keras.datasets.fashion_mnist.load_data()
#定义分类信息
class_names = ['T-shirt/top', 'Trouser', 'Pullover', 'Dress', 'Coat',
               'Sandal', 'Shirt', 'Sneaker', 'Bag', 'Ankle boot']
#格式化训练图像
```

```
train_images = train_images / 255.0
test_images = test_images / 255.0
#构建模型
model = keras.Sequential(
#使用时序模型构建模型
[
    layers.Flatten(input_shape=[28, 28]),
#全连接层
    layers.Dense(128, activation='relu'),
    layers.Dense(10, activation='softmax')
])
#初始化模型参数
model.compile(optimizer='adam', loss='sparse_categorical_crossentropy',
    metrics=['accuracy'])
#输出模型详情
model.summary()
```

（11）代码的运行结果如下。

```
Model: " sequential "
_____
Layer (type)                 Output Shape              Param #
=================================================================
flatten (Flatten)            (None, 784)               0
_____
dense (Dense)                (None, 128)               100480
_____
dense_1 (Dense)              (None, 10)                1290
=================================================================
Total params: 101,770
Trainable params: 101,770
Non-trainable params: 0
_____
```

（12）对构造的网络进行训练和验证，代码如下。

```
import ssl
#以 TensorFlow 为基础构建 Keras
from tensorflow import keras
from tensorflow.keras import layers
#导入图形化工具
import matplotlib.pyplot as plt
ssl._create_default_https_context = ssl._create_unverified_context
#导入数据集并划分训练集和测试集
(train_images, train_labels), (test_images, test_labels) = keras.datasets.fashion_mnist.load_data()
#定义分类信息
```

```
class_names = ['T-shirt/top', 'Trouser', 'Pullover', 'Dress', 'Coat',
               'Sandal', 'Shirt', 'Sneaker', 'Bag', 'Ankle boot']
#格式化训练图像
train_images = train_images / 255.0
test_images = test_images / 255.0
#构建模型
model = keras.Sequential(
#使用时序模型构建模型
[
    layers.Flatten(input_shape=[28, 28]),
#全连接层
    layers.Dense(128, activation='relu'),
    layers.Dense(10, activation='softmax')
])
#初始化模型参数
model.compile(optimizer='adam', loss='sparse_categorical_crossentropy',
    metrics=['accuracy'])
#制订训练计划并执行，训练5次
model.fit(train_images, train_labels, epochs=5)
```

（13）代码的运行结果如下。

```
Epoch 1/5
60000/60000 [==============] - 5s 80us/sample - loss: 0.4998 - accuracy: 0.8242
...
Epoch 5/5
60000/60000 [==============] - 4s 64us/sample - loss: 0.2979 - accuracy: 0.8912
```

（14）在进行了多次训练后，模型的准确率得到了保证，可以使用模型对数据进行验证。选取一个图像进行验证，代码如下。

```
import ssl
#以 TensorFlow 为基础构建 Keras
from tensorflow import keras
from tensorflow.keras import layers
#导入图形化工具
import matplotlib.pyplot as plt
ssl._create_default_https_context = ssl._create_unverified_context
#导入数据集并划分训练集和测试集
(train_images, train_labels), (test_images, test_labels) = keras.datasets.fashion_mnist.load_data()
#定义分类信息
class_names = ['T-shirt/top', 'Trouser', 'Pullover', 'Dress', 'Coat',
               'Sandal', 'Shirt', 'Sneaker', 'Bag', 'Ankle boot']
#格式化训练图像
train_images = train_images / 255.0
```

```python
test_images = test_images / 255.0
#构建模型
model = keras.Sequential(
#使用时序模型构建模型
[
    layers.Flatten(input_shape=[28, 28]),
#全连接层
    layers.Dense(128, activation='relu'),
    layers.Dense(10, activation='softmax')
])
#初始化模型参数
model.compile(optimizer='adam', loss='sparse_categorical_crossentropy',
    metrics=['accuracy'])
#制订训练计划并执行,训练5次
model.fit(train_images, train_labels, epochs=5)
#输入测试图像,进行分类测试
predictions = model.predict(test_images)
#定义显示函数
def plot_image(i, predictions_array, true_label, img):
    predictions_array, true_label, img = predictions_array[i], true_label[i], img[i]
    plt.grid(False)
    plt.xticks([])
    plt.yticks([])
    plt.imshow(img, cmap=plt.cm.binary)
#定义分类中值最大的元素对应的索引
    predicted_label = np.argmax(predictions_array)
    if predicted_label == true_label:
        color = 'blue'
    else:
        color = 'red'
    plt.xlabel("{} {:2.0f}% ({})".format(class_names[predicted_label],
        100*np.max(predictions_array), class_names[true_label]), color=color)
def plot_value_array(i, predictions_array, true_label):
    predictions_array, true_label = predictions_array[i], true_label[i]
    plt.grid(False)
    plt.xticks([])
    plt.yticks([])
    thisplot = plt.bar(range(10), predictions_array, color="#777777")
    plt.ylim([0, 1])
    predicted_label = np.argmax(predictions_array)
    thisplot[predicted_label].set_color('red')
    thisplot[true_label].set_color('blue')
i = 0
```

```
plt.figure(figsize=(6,3))
plt.subplot(1,2,1)
plot_image(i, predictions, test_labels, test_images)
plt.subplot(1,2,2)
plot_value_array(i, predictions, test_labels)
plt.show()
```

(15)运行代码,得到预测结果如图 9-4 所示,可以看出,已经使用训练模型对图像进行了分类,本例中对验证图像的分类预测准确率为 84%。

ankle boot 84%(Ankle boot)
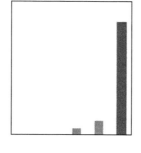

图 9-4　预测结果

(16)取多个图像进行批量验证,本例中取 15 个图像,代码如下。

```
import ssl
#以 TensorFlow 为基础构建 Keras
from tensorflow import keras
from tensorflow.keras import layers
#导入图形化工具
import matplotlib.pyplot as plt
ssl._create_default_https_context = ssl._create_unverified_context
#导入数据集并划分训练集和测试集
(train_images, train_labels), (test_images, test_labels) = keras.datasets.fashion_mnist.load_data()
#定义分类信息
class_names = ['T-shirt/top', 'Trouser', 'Pullover', 'Dress', 'Coat',
               'Sandal', 'Shirt', 'Sneaker', 'Bag', 'Ankle boot']
#格式化训练图像
train_images = train_images / 255.0
test_images = test_images / 255.0
#构建模型
model = keras.Sequential(
#使用时序模型构建模型
[
    layers.Flatten(input_shape=[28,28]),
#全连接层
```

```python
    layers.Dense(128, activation='relu'),
    layers.Dense(10, activation='softmax')
])
#初始化模型参数
model.compile(optimizer='adam', loss='sparse_categorical_crossentropy',
    metrics=['accuracy'])
#制订训练计划并执行,训练5次
model.fit(train_images, train_labels, epochs=5)
#输入测试图像,进行分类测试
predictions = model.predict(test_images)
def plot_image(i, predictions_array, true_label, img):
    predictions_array, true_label, img = predictions_array[i], true_label[i], img[i]
    plt.grid(False)
    plt.xticks([])
    plt.yticks([])
    plt.imshow(img, cmap=plt.cm.binary)
    predicted_label = np.argmax(predictions_array)
    if predicted_label == true_label:
        color = 'blue'
    else:
        color = 'red'
    plt.xlabel("{} {:2.0f}% ({})".format(class_names[predicted_label],
        100*np.max(predictions_array), class_names[true_label]), color=color)
def plot_value_array(i, predictions_array, true_label):
    predictions_array, true_label = predictions_array[i], true_label[i]
    plt.grid(False)
    plt.xticks([])
    plt.yticks([])
    thisplot = plt.bar(range(10), predictions_array, color="#777777")
    plt.ylim([0, 1])
    predicted_label = np.argmax(predictions_array)
    thisplot[predicted_label].set_color('red')
    thisplot[true_label].set_color('blue')
num_rows = 5
num_cols = 3
num_images = num_rows*num_cols
plt.figure(figsize=(2*2*num_cols, 2*num_rows))
for i in range(num_images):
    plt.subplot(num_rows, 2*num_cols, 2*i+1)
    plot_image(i, predictions, test_labels, test_images)
    plt.subplot(num_rows, 2*num_cols, 2*i+2)
    plot_value_array(i, predictions, test_labels)
plt.show()
```

(17)运行代码,得到多个图像的预测结果如图9-5所示,从图中可以看出,模型对某些类别的识别准确率不高,可以通过增加训练次数来提高识别的准确率。

图 9-5　多个图像的预测结果

9.2　图像识别

图像识别是机器学习的重要应用之一,其本质依然是使用大量的数据训练模型,以模拟人类神经的分辨能力并识别图像。

本节使用卷积神经网络对 MNIST 手写数字集进行识别。

(1) 使用 Keras.layers 提供的 Conv2D(卷积)与 MaxPooling2D(池化)函数构建一个基于卷积神经网络的模型,代码如下。

```
#以 TensorFlow 为基础构建 Keras
import tensorflow as tf
from tensorflow.keras import datasets, layers, models
#定义 CNN 模型类
class CNN(object):
    def __init__(self):
        model = models.Sequential()
        #第 1 层卷积,卷积核大小为 3*3,32 个,28*28 为待训练图像的大小
        model.add(layers.Conv2D(
```

```
            32, (3, 3), activation='relu', input_shape=(28, 28, 1)))
        model.add(layers.MaxPooling2D((2, 2)))
        #第2层卷积，卷积核大小为 3*3, 64 个
        model.add(layers.Conv2D(64, (3, 3), activation='relu'))
        model.add(layers.MaxPooling2D((2, 2)))
        #第3层卷积，卷积核大小为 3*3, 64 个
        model.add(layers.Conv2D(64, (3, 3), activation='relu'))
        model.add(layers.Flatten())
        #全连接层
        model.add(layers.Dense(64, activation='relu'))
        model.add(layers.Dense(10, activation='softmax'))
        #输出模型详情
        model.summary()
        self.model = model
#定义 Main()函数
if __name__ == "__main__":
    #调用自定义的 CNN 类
    CNN()
```

（2）运行代码，得到相应的网络结构，结果如下。

```
Model: "sequential"

_____
Layer (type)                 Output Shape              Param #
=================================================================
conv2d (Conv2D)              (None, 26, 26, 32)        320
_____
max_pooling2d (MaxPooling2D) (None, 13, 13, 32)        0
_____
conv2d_1 (Conv2D)            (None, 11, 11, 64)        18496
_____
max_pooling2d_1 (MaxPooling2 (None, 5, 5, 64)          0
_____
conv2d_2 (Conv2D)            (None, 3, 3, 64)          36928
_____
flatten (Flatten)            (None, 576)               0
_____
dense (Dense)                (None, 64)                36928
_____
dense_1 (Dense)              (None, 10)                650
=================================================================
Total params: 93,322
Trainable params: 93,322
Non-trainable params: 0
_____
```

从结果中可以看出，每个 Conv2D 和 MaxPooling2D 层的输出都是三维的张量(height, width,channel)。height 和 width 会逐渐变小。输出的 channel 的个数由第 1 个参数 height（如 32 或 64）控制。随着 height 和 width 变小，channel 可能变大。

（3）模型构建完成后，需要构造数据集，并在训练后输出相应的训练结果，代码如下。

```python
import ssl
import os
#以 TensorFlow 为基础构建 Keras
import tensorflow as tf
from tensorflow.keras import datasets, layers, models
ssl._create_default_https_context = ssl._create_unverified_context
#定义 CNN 模型类
class CNN(object):
    def __init__(self):
        model = models.Sequential()
        #第 1 层卷积，卷积核大小为 3*3，32 个，28*28 为待训练图像的大小
        model.add(layers.Conv2D(
            32, (3, 3), activation='relu', input_shape=(28, 28, 1)))
        model.add(layers.MaxPooling2D((2, 2)))
        #第 2 层卷积，卷积核大小为 3*3，64 个
        model.add(layers.Conv2D(64, (3, 3), activation='relu'))
        model.add(layers.MaxPooling2D((2, 2)))
        #第 3 层卷积，卷积核大小为 3*3，64 个
        model.add(layers.Conv2D(64, (3, 3), activation='relu'))
        model.add(layers.Flatten())
        #全连接层
        model.add(layers.Dense(64, activation='relu'))
        model.add(layers.Dense(10, activation='softmax'))
        #输出模型详情
        model.summary()
        self.model = model
#定义数据源处理类
class DataSource(object):
    def __init__(self):
        #导入 MNIST 数据集
        (train_images, train_labels), (test_images,
            test_labels) = datasets.mnist.load_data()
        #60000 张训练图片，10000 张测试图片
        train_images = train_images.reshape((60000, 28, 28, 1))
        test_images = test_images.reshape((10000, 28, 28, 1))
        #将像素映射到 0~1
        train_images, test_images = train_images / 255.0, test_images / 255.0
        self.train_images, self.train_labels = train_images, train_labels
        self.test_images, self.test_labels = test_images, test_labels
```

```python
#定义训练类
class Train:
    def __init__(self):
        self.cnn = CNN()
        self.data = DataSource()
    def train(self):
        check_path = './ckpt/cp-{epoch:04d}.ckpt'
        #定义回调函数,每训练5次保存一次检查点
        save_model_cb = tf.keras.callbacks.ModelCheckpoint(
            check_path, save_weights_only=True, verbose=1, period=5)
        #定义模型参数
        self.cnn.model.compile(optimizer='adam',
            loss='sparse_categorical_crossentropy', metrics=['accuracy'])
        #制订训练计划
        self.cnn.model.fit(self.data.train_images, self.data.train_labels,
            epochs=5, callbacks=[save_model_cb])
        #对模型进行评估
        test_loss, test_acc = self.cnn.model.evaluate(
            self.data.test_images, self.data.test_labels)
        print("准确率: %.4f,共测试了%d 张图片 " % (test_acc, len(self.data.test_labels)))
#定义Main()函数
if __name__ == "__main__":
    #进行训练
    test = Train()
    test.train()
```

(4) 代码的运行结果如下。

```
Model: " sequential "
_____
Layer (type)                 Output Shape              Param #
=================================================================
conv2d (Conv2D)              (None, 26, 26, 32)        320
_____
max_pooling2d (MaxPooling2D) (None, 13, 13, 32)        0
_____
conv2d_1 (Conv2D)            (None, 11, 11, 64)        18496
_____
max_pooling2d_1 (MaxPooling2 (None, 5, 5, 64)          0
_____
conv2d_2 (Conv2D)            (None, 3, 3, 64)          36928
_____
flatten (Flatten)            (None, 576)               0
```

```
_____
dense (Dense)                (None, 64)                36928
_____
dense_1 (Dense)              (None, 10)                650
=================================================================
Total params: 93,322
Trainable params: 93,322
Non-trainable params: 0
_____

Train on 60000 samples
Epoch 1/5
60000/60000 [==============] - 58s 968us/sample - loss: 0.1491 - accuracy: 0.9540
...
Epoch 5/5
59936/60000 [===========>.] - ETA: 0s - loss: 0.0195 - accuracy: 0.9940
Epoch 00005: saving model to ./ckpt/cp-0005.ckpt
60000/60000 [==============] - 51s 849us/sample - loss: 0.0195 - accuracy: 0.9940
10000/1 [==============] - 2s 228us/sample - loss: 0.0135 - accuracy: 0.9913
准确率: 0.9913, 共测试了 10000 张图片
```

在第一轮训练后，准确率达到了 0.9540，训练 5 次后使用测试集进行验证，准确率达到 0.9913。在第 5 轮训练后，成功将模型参数保存在 ./ckpt/cp-0005.ckpt 中。

（5）多次训练保证了模型训练的准确性，使用保存的模型进行图像识别，代码如下。

```
from PIL import Image
#导入 NumPy 模块
import numpy as np
#以 TensorFlow 为基础构建 Keras
import tensorflow as tf
from tensorflow.keras import datasets, layers, models
#定义 CNN 模型类
class CNN(object):
    def __init__(self):
        model = models.Sequential()
        #第 1 层卷积，卷积核大小为 3*3, 32 个，28*28 为待训练图片的大小
        model.add(layers.Conv2D(
            32, (3, 3), activation='relu', input_shape=(28, 28, 1)))
        model.add(layers.MaxPooling2D((2, 2)))
        #第 2 层卷积，卷积核大小为 3*3, 64 个
        model.add(layers.Conv2D(64, (3, 3), activation='relu'))
        model.add(layers.MaxPooling2D((2, 2)))
        #第 3 层卷积，卷积核大小为 3*3, 64 个
        model.add(layers.Conv2D(64, (3, 3), activation='relu'))
```

```python
        model.add(layers.Flatten())
#全连接层
        model.add(layers.Dense(64, activation='relu'))
        model.add(layers.Dense(10, activation='softmax'))
#输出模型详情
        model.summary()
        self.model = model
#定义预测类
class Predict(object):
    def __init__(self):
        latest = tf.train.latest_checkpoint('./ckpt')
        self.cnn = CNN()
        #恢复网络权重
        self.cnn.model.load_weights(latest)
    def predict(self, image_path):
        #以黑白方式读取图片
        img = Image.open(image_path).convert('L')
        flatten_img = np.reshape(img, (28, 28, 1))
        x = np.array([1 - flatten_img])
        y = self.cnn.model.predict(x)
        #因为 x 只传入了一张图片，取 y[0]即可
        #np.argmax()取得最大值的下标，即代表的数字
        print(image_path)
        print(y[0])
        print('          -> Predict digit', np.argmax(y[0]))
if __name__ == "__main__":
    test = Predict()
    #对测试图片进行预测
    test.predict('./image_test/test_0.png')
    test.predict('./image_test/test_1.png')
    test.predict('./image_test/test_4.png')
```

（6）本例使用图 9-6、图 9-7 和图 9-8 进行图像识别。

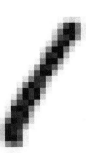

图 9-6 test_0　　　　　　图 9-7 test_1　　　　　　图 9-8 test_4

（7）代码的运行结果如下。

```
Model: " sequential "
_____
Layer (type)                 Output Shape              Param #
=================================================================
conv2d (Conv2D)              (None, 26, 26, 32)        320
_____
max_pooling2d (MaxPooling2D) (None, 13, 13, 32)        0
_____
conv2d_1 (Conv2D)            (None, 11, 11, 64)        18496
_____
max_pooling2d_1 (MaxPooling2 (None, 5, 5, 64)          0
_____
conv2d_2 (Conv2D)            (None, 3, 3, 64)          36928
_____
flatten (Flatten)            (None, 576)               0
_____
dense (Dense)                (None, 64)                36928
_____
dense_1 (Dense)              (None, 10)                650
=================================================================
Total params: 93,322
Trainable params: 93,322
Non-trainable params: 0
_____
./image_test/test_0.png
[1. 0. 0. 0. 0. 0. 0. 0. 0. 0.]
        -> Predict digit 0
./image_test/test_1.png
[0. 1. 0. 0. 0. 0. 0. 0. 0. 0.]
        -> Predict digit 1
./image_test/test_4.png
[0. 0. 0. 0. 1. 0. 0. 0. 0. 0.]
        -> Predict digit 4
```

从结果可以看出，经过机器学习，模型具备了基本的图像识别功能，并准确识别了示例图像。

9.3 生成对抗网络

生成对抗网络（Generative Adversarial Network，GAN）是一类功能强大、应用广泛的神经网络。由两个神经网络组成：生成器和判别器，这两个网络相互制约。通过这种制约，能够学习几乎所有类型的数据分布。生成对抗网络的流程图如图9-9所示。

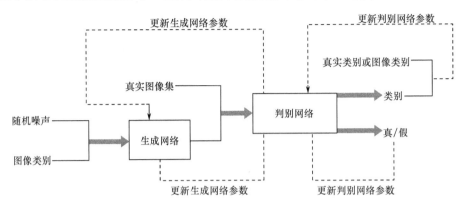

图9-9 生成对抗网络的流程图

在了解生成对抗网络的基本概念后，介绍如何使用生成对抗网络生成图像样本。

（1）设定参数并导入数据集，代码如下。

```
from __future__ import absolute_import, print_function, division
import tensorflow as tf
import tensorflow.keras as keras
import matplotlib.pyplot as plt
import numpy as np
import os
import PIL
import imageio
import glob
import time
#定义超参
BUFFER_SIZE = 60000
BATCH_SIZE = 256
EPOCHS = 50
z_dim = 100
num_examples_to_generate = 16
seed = tf.random.normal([num_examples_to_generate, z_dim])
#导入数据集
(train_images, train_labels), (_, _) = keras.datasets.mnist.load_data()
#显示数据集图像
plt.imshow(train_images[0])
plt.show()
```

（2）运行代码，得到测试数据集图像如图9-10所示。

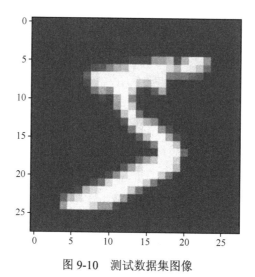

图 9-10 测试数据集图像

(3) 构建一个生成器,本例使用 Keras 中的序列模型构建简单的卷积网络模型,代码如下。

```
from __future__ import absolute_import, print_function, division
import tensorflow as tf
import tensorflow.keras as keras
import matplotlib.pyplot as plt
import numpy as np
import os
import PIL
import imageio
import glob
import time
#定义超参
BUFFER_SIZE = 60000
BATCH_SIZE = 256
EPOCHS = 50
z_dim = 100
num_examples_to_generate = 16
seed = tf.random.normal([num_examples_to_generate, z_dim])
#导入数据集
(train_images, train_labels), (_, _) = keras.datasets.mnist.load_data()
#对数据进行处理
train_images = train_images.reshape(train_images.shape[0], 28, 28, 1).astype('float32')
train_images = (train_images - 127.5) / 127.5
train_dataset = tf.data.Dataset.from_tensor_slices(train_images).shuffle(BUFFER_SIZE).batch(BATCH_SIZE)
#构建生成器
def make_generator():
    generator = keras.Sequential([
```

```
            keras.layers.Dense(7 * 7 * 256, use_bias=False, input_shape=(100,)),
            keras.layers.BatchNormalization(),
            keras.layers.LeakyReLU(),
            keras.layers.Reshape((7, 7, 256)),
            keras.layers.Conv2DTranspose(128, (5, 5), strides=(1, 1), padding='same', use_bias=False),
            keras.layers.BatchNormalization(),
            keras.layers.LeakyReLU(),
            keras.layers.Conv2DTranspose(64, (5, 5), strides=(2, 2), padding='same', use_bias=False),
            keras.layers.BatchNormalization(),
            keras.layers.LeakyReLU(),
            keras.layers.Conv2DTranspose(1, (5, 5), strides=(2, 2), padding='same', use_bias=False, activation='tanh'),
        ])
        return generator
#对生成器进行测试
g = make_generator()
z = tf.random.normal([1, 100])
fake_image = g(z, training=False)
plt.imshow(fake_image[0, :, :, 0], cmap='gray')
plt.show()
```

(4) 运行代码,得到生成器的验证结果如图 9-11 所示。

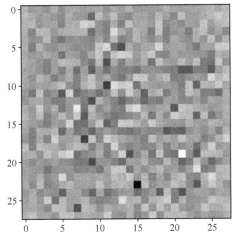

图 9-11 生成器的验证结果

(5) 构建一个判别器并测试,代码如下。

```
from __future__ import absolute_import, print_function, division
import tensorflow as tf
import tensorflow.keras as keras
import matplotlib.pyplot as plt
```

```python
import numpy as np
import os
import PIL
import imageio
import glob
import time
#定义超参
BUFFER_SIZE = 60000
BATCH_SIZE = 256
EPOCHS = 50
z_dim = 100
num_examples_to_generate = 16
seed = tf.random.normal([num_examples_to_generate, z_dim])
#导入数据集
(train_images, train_labels), (_, _) = keras.datasets.mnist.load_data()
#对数据集进行处理
train_images = train_images.reshape(train_images.shape[0], 28, 28, 1).astype('float32')
train_images = (train_images - 127.5) / 127.5
train_dataset = tf.data.Dataset.from_tensor_slices(train_images).shuffle(BUFFER_SIZE).batch(BATCH_SIZE)
#构建生成器
def make_generator():
    generator = keras.Sequential([
        keras.layers.Dense(7 * 7 * 256, use_bias=False, input_shape=(100,)),
        keras.layers.BatchNormalization(),
        keras.layers.LeakyReLU(),
        keras.layers.Reshape((7, 7, 256)),
        keras.layers.Conv2DTranspose(128, (5, 5), strides=(1, 1), padding='same', use_bias=False),
        keras.layers.BatchNormalization(),
        keras.layers.LeakyReLU(),
        keras.layers.Conv2DTranspose(64, (5, 5), strides=(2, 2), padding='same', use_bias=False),
        keras.layers.BatchNormalization(),
        keras.layers.LeakyReLU(),
        keras.layers.Conv2DTranspose(1, (5, 5), strides=(2, 2), padding='same', use_bias=False, activation='tanh'),
    ])
    return generator
#构建判别器
def make_discriminator():
    discriminator = keras.Sequential([
        keras.layers.Conv2D(64, (5, 5), strides=(2, 2), padding='same'),
        keras.layers.LeakyReLU(),
```

```
                keras.layers.Dropout(0.2),
                keras.layers.Conv2D(128, (5, 5), strides=(2, 2), padding='same'),
                keras.layers.LeakyReLU(),
                keras.layers.Dropout(0.2),
                keras.layers.Flatten(),
                keras.layers.Dense(1),
        ])
        return discriminator
#对判别器进行测试
g = make_generator()
z = tf.random.normal([1,100])
fake_image = g(z,training = False)
d = make_discriminator()
pred = d(fake_image)
print('pred score is: ', pred)
```

(6) 代码的运行结果如下。

```
pred score is:  tf.Tensor([[-0.00185565]], shape=(1, 1), dtype=float32)
```

(7) 定义损失函数及优化器，代码如下。

```
from __future__ import absolute_import, print_function, division
import tensorflow as tf
import tensorflow.keras as keras
import matplotlib.pyplot as plt
import numpy as np
import os
import PIL
import imageio
import glob
import time
#定义超参
BUFFER_SIZE = 60000
BATCH_SIZE = 256
EPOCHS = 50
z_dim = 100
num_examples_to_generate = 16
seed = tf.random.normal([num_examples_to_generate, z_dim])
#导入数据集
(train_images, train_labels), (_, _) = keras.datasets.mnist.load_data()
#对数据集进行处理
train_images = train_images.reshape(train_images.shape[0], 28, 28, 1).astype('float32')
train_images = (train_images - 127.5) / 127.5
train_dataset = tf.data.Dataset.from_tensor_slices(train_images).shuffle(BUFFER_SIZE).batch(BATCH_SIZE)
```

```python
#构建生成器
def make_generator():
    generator = keras.Sequential([
        keras.layers.Dense(7 * 7 * 256, use_bias=False, input_shape=(100,)),
        keras.layers.BatchNormalization(),
        keras.layers.LeakyReLU(),
        keras.layers.Reshape((7, 7, 256)),
        keras.layers.Conv2DTranspose(128, (5, 5), strides=(1, 1), padding='same', use_bias=False),
        keras.layers.BatchNormalization(),
        keras.layers.LeakyReLU(),
        keras.layers.Conv2DTranspose(64, (5, 5), strides=(2, 2), padding='same', use_bias=False),
        keras.layers.BatchNormalization(),
        keras.layers.LeakyReLU(),
        keras.layers.Conv2DTranspose(1, (5, 5), strides=(2, 2), padding='same', use_bias=False, activation='tanh'),
    ])
    return generator
#构建判别器
def make_discriminator():
    discriminator = keras.Sequential([
        keras.layers.Conv2D(64, (5, 5), strides=(2, 2), padding='same'),
        keras.layers.LeakyReLU(),
        keras.layers.Dropout(0.2),
        keras.layers.Conv2D(128, (5, 5), strides=(2, 2), padding='same'),
        keras.layers.LeakyReLU(),
        keras.layers.Dropout(0.2),
        keras.layers.Flatten(),
        keras.layers.Dense(1),
    ])
    return discriminator
#定义损失函数和优化器
cross_entropy = tf.keras.losses.BinaryCrossentropy(from_logits = True)
g_optimizer = keras.optimizers.Adam(1e-4)
d_optimizer = keras.optimizers.Adam(1e-4)
```

(8) 分别定义生成器和判别器的损失函数, 判别器的损失项包含两部分: 对生成图像的判别和对真实图像的判别, 代码如下。

```python
from __future__ import absolute_import, print_function, division
import tensorflow as tf
import tensorflow.keras as keras
import matplotlib.pyplot as plt
import numpy as np
```

```python
import os
import PIL
import imageio
import glob
import time
#定义超参
BUFFER_SIZE = 60000
BATCH_SIZE = 256
EPOCHS = 50
z_dim = 100
num_examples_to_generate = 16
seed = tf.random.normal([num_examples_to_generate, z_dim])
#导入数据集
(train_images, train_labels), (_, _) = keras.datasets.mnist.load_data()
#对数据集进行处理
train_images = train_images.reshape(train_images.shape[0], 28, 28, 1).astype('float32')
train_images = (train_images - 127.5) / 127.5
train_dataset = tf.data.Dataset.from_tensor_slices(train_images).shuffle(BUFFER_SIZE).batch(BATCH_SIZE)
#构建生成器
def make_generator():
    generator = keras.Sequential([
        keras.layers.Dense(7 * 7 * 256, use_bias=False, input_shape=(100,)),
        keras.layers.BatchNormalization(),
        keras.layers.LeakyReLU(),
        keras.layers.Reshape((7, 7, 256)),
        keras.layers.Conv2DTranspose(128, (5, 5), strides=(1, 1), padding='same', use_bias=False),
        keras.layers.BatchNormalization(),
        keras.layers.LeakyReLU(),
        keras.layers.Conv2DTranspose(64, (5, 5), strides=(2, 2), padding='same', use_bias=False),
        keras.layers.BatchNormalization(),
        keras.layers.LeakyReLU(),
        keras.layers.Conv2DTranspose(1, (5, 5), strides=(2, 2), padding='same', use_bias=False, activation='tanh'),
    ])
    return generator
#构建判别器
def make_discriminator():
    discriminator = keras.Sequential([
        keras.layers.Conv2D(64, (5, 5), strides=(2, 2), padding='same'),
```

```python
        keras.layers.LeakyReLU(),
        keras.layers.Dropout(0.2),
        keras.layers.Conv2D(128, (5, 5), strides=(2, 2), padding='same'),
        keras.layers.LeakyReLU(),
        keras.layers.Dropout(0.2),
        keras.layers.Flatten(),
        keras.layers.Dense(1),
    ])
    return discriminator
#定义损失函数和优化器
cross_entropy = tf.keras.losses.BinaryCrossentropy(from_logits = True)
g_optimizer = keras.optimizers.Adam(1e-4)
d_optimizer = keras.optimizers.Adam(1e-4)
#定义生成器和判别器的损失函数
def generator_loss(fake_iamge):
    return cross_entropy(tf.ones_like(fake_iamge),fake_iamge)
def discriminator_loss(fake_iamge,real_iamge):
    real_loss = cross_entropy(tf.ones_like(real_iamge),real_iamge)
    fake_loss = cross_entropy(tf.ones_like(fake_iamge),fake_iamge)
    return real_loss + fake_loss
```

（9）训练模型并显示图像，代码如下。

```python
from __future__ import absolute_import, print_function, division
import tensorflow as tf
import tensorflow.keras as keras
import matplotlib.pyplot as plt
import numpy as np
import os
import PIL
import imageio
import glob
import time
#定义超参
BUFFER_SIZE = 60000
BATCH_SIZE = 256
EPOCHS = 50
z_dim = 100
num_examples_to_generate = 16
seed = tf.random.normal([num_examples_to_generate, z_dim])
#导入数据集
(train_images, train_labels), (_, _) = keras.datasets.mnist.load_data()
#对数据集进行处理
train_images = train_images.reshape(train_images.shape[0], 28, 28, 1).astype('float32')
```

```python
    train_images = (train_images - 127.5) / 127.5
    train_dataset = tf.data.Dataset.from_tensor_slices(train_images).shuffle(BUFFER_SIZE).batch(BATCH_SIZE)
    #构建生成器
    def make_generator():
        generator = keras.Sequential([
            keras.layers.Dense(7 * 7 * 256, use_bias=False, input_shape=(100,)),
            keras.layers.BatchNormalization(),
            keras.layers.LeakyReLU(),
            keras.layers.Reshape((7, 7, 256)),
            keras.layers.Conv2DTranspose(128, (5, 5), strides=(1, 1), padding='same', use_bias=False),
            keras.layers.BatchNormalization(),
            keras.layers.LeakyReLU(),
            keras.layers.Conv2DTranspose(64, (5, 5), strides=(2, 2), padding='same', use_bias=False),
            keras.layers.BatchNormalization(),
            keras.layers.LeakyReLU(),
            keras.layers.Conv2DTranspose(1, (5, 5), strides=(2, 2), padding='same', use_bias=False, activation='tanh'),
        ])
        return generator
    g = make_generator()
    z = tf.random.normal([1, 100])
    fake_image = g(z, training=False)
    #构建判别器
    def make_discriminator():
        discriminator = keras.Sequential([
            keras.layers.Conv2D(64, (5, 5), strides=(2, 2), padding='same'),
            keras.layers.LeakyReLU(),
            keras.layers.Dropout(0.2),
            keras.layers.Conv2D(128, (5, 5), strides=(2, 2), padding='same'),
            keras.layers.LeakyReLU(),
            keras.layers.Dropout(0.2),
            keras.layers.Flatten(),
            keras.layers.Dense(1),
        ])
        return discriminator
    d = make_discriminator()
    pred = d(fake_image)
    #定义损失函数和优化器
    cross_entropy = tf.keras.losses.BinaryCrossentropy(from_logits=True)
    g_optimizer = keras.optimizers.Adam(1e-4)
```

```python
d_optimizer = keras.optimizers.Adam(1e-4)
#定义生成器和判别器的损失函数
def generator_loss(fake_iamge):
    return cross_entropy(tf.ones_like(fake_iamge), fake_iamge)
def discriminator_loss(fake_iamge, real_iamge):
    real_loss = cross_entropy(tf.ones_like(real_iamge), real_iamge)
    fake_loss = cross_entropy(tf.ones_like(fake_iamge), fake_iamge)
    return real_loss + fake_loss
#设置检查点
checkpoint_dir = './training_checkpoints'
checkpoint_prefix = os.path.join(checkpoint_dir, "ckpt")
checkpoint = tf.train.Checkpoint(g_optimizer=g_optimizer,
    d_optimizer=d_optimizer, g=g, d=d)
#定义大批数据训练过程，注意此处使用了 tf.fuction
@tf.function
def train_one_step(images):
    z = tf.random.normal([BATCH_SIZE, z_dim])
    with tf.GradientTape() as g_tape, tf.GradientTape() as d_tape:
        fake_images = g(z, training=True)
        real_pred = d(images, training=True)
        fake_pred = d(fake_images, training=True)
        g_loss = generator_loss(fake_images)
        d_loss = discriminator_loss(real_pred, fake_pred)
    g_gradients = g_tape.gradient(g_loss, g.trainable_variables)
    d_gradients = d_tape.gradient(d_loss, d.trainable_variables)
    g_optimizer.apply_gradients(zip(g_gradients, g.trainable_variables))
    d_optimizer.apply_gradients(zip(d_gradients, d.trainable_variables))
#整个数据集的训练过程
def train(dataset, epochs):
    for epoch in range(epochs):
        start = time.time()
        for image_batch in dataset:
            train_one_step(image_batch)
        generate_and_save_images(g, epoch + 1, seed)
        if (epoch + 1) % 15 == 0:
            checkpoint.save(file_prefix=checkpoint_prefix)
        print('Time for epoch {} is {} sec'.format(epoch + 1, time.time() - start))
    generate_and_save_images(g, epochs, seed)
#显示图像
def generate_and_save_images(model, epoch, test_input):
    predictions = model(test_input, training=False)
    fig = plt.figure(figsize=(4, 4))
    for i in range(predictions.shape[0]):
        plt.subplot(4, 4, i + 1)
```

```
            plt.imshow(predictions[i, :, :, 0] * 127.5 + 127.5, cmap='gray')
            plt.axis('off')
        plt.savefig('image_at_epoch_{:04d}.png'.format(epoch))
        plt.show()
    if __name__ == '__main__':
        train(train_dataset, EPOCHS)
```

说明：tf.function 装饰器可以将代码转换为 TensorFlow 2.0 的图。

（10）运行代码，得到对抗网络的示例如图 9-12 所示。

本章对生成对抗网络进行了简单的介绍，建议在练习时使用多个 GPU 计算并运行代码，结果用 GIF 图像显示。

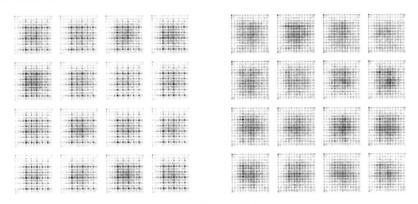

图 9-12　对抗网络的示例

第 10 章　TensorFlow 决策树

本章使用决策树和 tf.estimator API 训练 Boosted Trees 模型。Boosted Trees 模型是回归分析和分类任务中最受欢迎且最有效的机器学习方法之一，是一种集合技术。

10.1　Boosted Trees 简介

Boosted Trees 不断根据特定的特征添加分支，形成类似树状的数据结构，再形成一个多分支的树状结构模型，并将最终的分支节点作为评分标准来进行评分。将所有的分支节点相加即可得到预测值。

10.2　数据预测

Boosted Trees 模型可以通过微调超参来提升性能，本节以泰坦尼克数据集为例进行分析。
（1）加载泰坦尼克数据集，代码如下。

```
from __future__ import absolute_import, division, print_function
#导入 NumPy 模块
import numpy as np
import ssl
#导入 pandas 模块
import pandas as pd
ssl._create_default_https_context = ssl._create_unverified_context
#下载后缀为 csv 的数据集，并将其作为训练集和测试集
df_train = pd.read_csv('https://storage.googleapis.com/tf-datasets/titanic/train.csv')
df_test = pd.read_csv('https://storage.googleapis.com/tf-datasets/titanic/eval.csv')
```

说明：train.csv 为训练集，用于构建与生存相关的模型；test.csv 为测试集，用于验证模型。
（2）查看前 5 行数据，代码如下。

```
from __future__ import absolute_import, division, print_function
#导入 NumPy 模块
import numpy as np
import ssl
#导入 pandas 模块
import pandas as pd
ssl._create_default_https_context = ssl._create_unverified_context
```

```
#下载后缀为csv的数据集,并将其作为训练集和测试集
df_train = pd.read_csv('https://storage.googleapis.com/tf-datasets/titanic/train.csv')
df_test = pd.read_csv('https://storage.googleapis.com/tf-datasets/titanic/eval.csv')
#查看前5行数据
head = df_train.head()
print (head)
```

说明:pandas是一种基于NumPy的工具,大量应用于操作大型数据集。

(3)代码的运行结果如下。

```
   survived  sex     age   n_siblings_spouses  parch  fare     class   deck     embark_town  alone
0  0         male    22.0  1                   0      7.2500   Third   unknown  Southampton  n
1  1         female  38.0  1                   0      71.2833  First   C        Cherbourg    n
2  1         female  26.0  0                   0      7.9250   Third   unknown  Southampton  y
3  1         female  35.0  1                   0      53.1000  First   C        Southampton  n
4  0         male    28.0  0                   0      8.4583   Third   unknown  Queenstown   y

5 rows x 10 columns]
```

(4)查看后5行数据,代码如下。

```
from __future__ import absolute_import, division, print_function
#导入NumPy模块
import numpy as np
import ssl
#导入pandas模块
import pandas as pd
ssl._create_default_https_context = ssl._create_unverified_context
#下载后缀为csv的数据集,并将其作为训练集和测试集
df_train = pd.read_csv('https://storage.googleapis.com/tf-datasets/titanic/train.csv')
df_test = pd.read_csv('https://storage.googleapis.com/tf-datasets/titanic/eval.csv')
#查看后5行数据
tail = df_train.tail()
print (tail)
```

(5)代码的运行结果如下。

```
     survived  sex     age   n_siblings_spouses  parch  fare    class    deck     embark_town  alone
622  0         male    28.0  0                   0      10.50   Second   unknown  Southampton  y
623  0         male    25.0  0                   0      7.05    Third    unknown  Southampton  y
624  1         female  19.0  0                   0      30.00   First    B        Southampton  y
625  0         female  28.0  1                   2      23.45   Third    unknown  Southampton  n
626  0         male    32.0  0                   0      7.75    Third    unknown  Queenstown   y
```

(6)可以发现共有627行数据,数据字段对应表如表10-1所示。

表 10-1 数据字段对应表

属性名称	属性描述	属性名称	属性描述
sex	乘客性别	class	船舱等级
age	乘客年龄	deck	甲板编号
n_siblings_spouses	随行兄弟或者配偶数量	embark_town	登船地点
parch	随行父母或者子女数量	alone	是否为独自旅行
fare	费用		

（7）在确定相应字段的情况下，查看数据表的整体信息，代码如下。

```
from __future__ import absolute_import, division, print_function
#导入 NumPy 模块
import numpy as np
import ssl
#导入 pandas 模块
import pandas as pd
ssl._create_default_https_context = ssl._create_unverified_context
#下载后缀为 csv 的数据集，并将其作为训练集和测试集
df_train = pd.read_csv('https://storage.googleapis.com/tf-datasets/titanic/train.csv')
df_test = pd.read_csv('https://storage.googleapis.com/tf-datasets/titanic/eval.csv')
#查看数据信息，包含数据维度、数据类型、所占空间等信息
info = df_train.info()
print (info)
```

（8）代码的运行结果如下。

```
<class 'pandas.core.frame.DataFrame'>
RangeIndex: 627 entries, 0 to 626
Data columns (total 10 columns):
survived              627 non-null int64
sex                   627 non-null object
age                   627 non-null float64
n_siblings_spouses    627 non-null int64
parch                 627 non-null int64
fare                  627 non-null float64
class                 627 non-null object
deck                  627 non-null object
embark_town           627 non-null object
alone                 627 non-null object
dtypes: float64(2), int64(3), object(5)
memory usage: 49.1+ KB
None
```

从结果中可以看出，数据维度为 627 行×12 列，数据类型有 2 个 64 位的浮点型数据、3 个 64 位的整形数据、5 个 Python 对象数据。

(9)对数据集进行描述性统计,代码如下。

```
from __future__ import absolute_import, division, print_function
#导入 NumPy 模块
import numpy as np
import ssl
#导入 pandas 模块
import pandas as pd
ssl._create_default_https_context = ssl._create_unverified_context
#下载后缀为 csv 的数据集,并将其作为训练集和测试集
df_train = pd.read_csv('https://storage.googleapis.com/tf-datasets/titanic/train.csv')
df_test = pd.read_csv('https://storage.googleapis.com/tf-datasets/titanic/eval.csv')
#查看数据集描述性统计
describe = df_train.describe()
print (describe)
```

(10)代码的运行结果如下。

	survived	age	n_siblings_spouses	parch	fare
count	627.000000	627.000000	627.000000	627.000000	627.000000
mean	0.387560	29.631308	0.545455	0.379585	34.385399
std	0.487582	12.511818	1.151090	0.792999	54.597730
min	0.000000	0.750000	0.000000	0.000000	0.000000
25%	0.000000	23.000000	0.000000	0.000000	7.895800
50%	0.000000	28.000000	0.000000	0.000000	15.045800
75%	1.000000	35.000000	1.000000	0.000000	31.387500
max	1.000000	80.000000	8.000000	5.000000	512.329200

可以看出,除了 Python 对象,其他数据类型均参与了计算。只有 38.7560%的人幸存,死亡率很高。

(11)在了解每个特征的大致信息的情况下,对特征进行分析,下面是显示年龄分布的代码。

```
from __future__ import absolute_import, division, print_function
#导入 NumPy 模块
import numpy as np
import ssl
#导入 pandas 模块
import pandas as pd
#导入图形化工具
import matplotlib.pyplot as plt
ssl._create_default_https_context = ssl._create_unverified_context
#下载后缀为 csv 的数据集,并将其作为训练集和测试集
df_train = pd.read_csv('https://storage.googleapis.com/tf-datasets/titanic/train.csv')
```

```
df_test = pd.read_csv('https://storage.googleapis.com/tf-datasets/titanic/eval.csv')
#显示年龄分布
df_train.age.hist(bins=20)
plt.show()
```

（12）运行代码，得到乘客的年龄分布如图 10-1 所示。

由图 10-1 可知，20～30 岁的乘客居多。

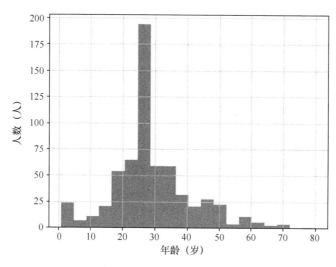

图 10-1　乘客的年龄分布

（13）显示性别分布，代码如下。

```
from __future__ import absolute_import, division, print_function
#导入 NumPy 模块
import numpy as np
import ssl
#导入 pandas 模块
import pandas as pd
#导入图形化工具
import matplotlib.pyplot as plt
ssl._create_default_https_context = ssl._create_unverified_context
#下载后缀为 csv 的数据集，并将其作为训练集和测试集
df_train = pd.read_csv('https://storage.googleapis.com/tf-datasets/titanic/train.csv')
df_test = pd.read_csv('https://storage.googleapis.com/tf-datasets/titanic/eval.csv')
#显示性别分布
df_train.sex.value_counts().plot(kind='barh')
plt.show()
```

（14）运行代码，得到乘客的性别分布如图 10-2 所示。

由图 10-2 可知，男性乘客的数量几乎是女性乘客数量的两倍。

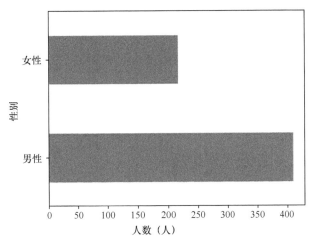

图 10-2　乘客的性别分布

（15）显示船舱等级分布，这个参数间接体现了乘客的经济实力，代码如下。

```
from __future__ import absolute_import, division, print_function
#导入 NumPy 模块
import numpy as np
import ssl
#导入 pandas 模块
import pandas as pd
#导入图形化工具
import matplotlib.pyplot as plt
ssl._create_default_https_context = ssl._create_unverified_context
#下载后缀为 csv 的数据集，并将其作为训练集和测试集
df_train = pd.read_csv('https://storage.googleapis.com/tf-datasets/titanic/train.csv')
df_test = pd.read_csv('https://storage.googleapis.com/tf-datasets/titanic/eval.csv')
#显示船舱等级分布
df_train['class'].value_counts().plot(kind='barh')
plt.show()
```

（16）运行代码，得到船舱等级分布如图 10-3 所示。

由图 10-3 可知，三等舱乘客占比较高。

（17）显示上船地点分布，代码如下。

```
from __future__ import absolute_import, division, print_function
#导入 NumPy 模块
import numpy as np
import ssl
#导入 pandas 模块
import pandas as pd
#导入图形化工具
import matplotlib.pyplot as plt
ssl._create_default_https_context = ssl._create_unverified_context
```

```
#下载后缀为csv的数据集,并将其作为训练集和测试集
df_train = pd.read_csv('https://storage.googleapis.com/tf-datasets/titanic/train.csv')
df_test = pd.read_csv('https://storage.googleapis.com/tf-datasets/titanic/eval.csv')
#显示上船地点分布
df_train['embark_town'].value_counts().plot(kind='barh')
plt.show()
```

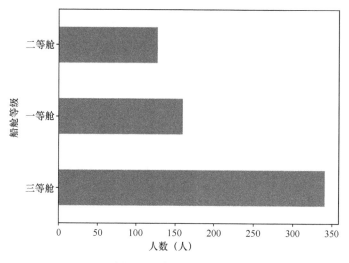

图 10-3 船舱等级分布

(18) 运行代码,得到上船地点分布如图 10-4 所示。

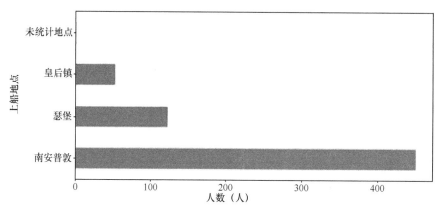

图 10-4 上船地点分布

由图 10-4 可知,大部分乘客从南安普敦上船。
(19) 对相关的特征和存活率进行关联计算,显示性别和存活率的关系。

```
from __future__ import absolute_import, division, print_function
#导入NumPy模块
import numpy as np
import ssl
#导入pandas模块
```

```
import pandas as pd
#导入图形化工具
import matplotlib.pyplot as plt
ssl._create_default_https_context = ssl._create_unverified_context
#下载后缀为 csv 的数据集,并将其作为训练集和测试集
df_train = pd.read_csv('https://storage.googleapis.com/tf-datasets/titanic/train.csv')
df_test = pd.read_csv('https://storage.googleapis.com/tf-datasets/titanic/eval.csv')
#分离标注字段
y_train = df_train.pop('survived')
y_test = df_test.pop('survived')
#显示性别和存活率的关系
pd.concat([df_train, y_train], axis=1).groupby('sex').survived.mean().plot(kind='barh').set_xlabel('% survive')
plt.show()
```

(20) 运行代码,得到性别和存活率的关系如图 10-5 所示。

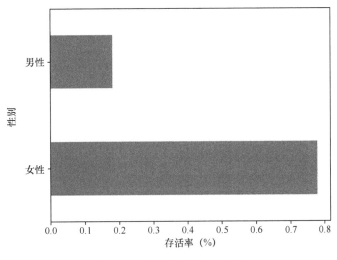

图 10-5　性别和存活率的关系

由图 10-5 可知,女性的存活率较高,几乎为男性的 5 倍。

(21) 对年龄进行分段,统计不同年龄的乘客的存活率,代码如下。

```
from __future__ import absolute_import, division, print_function
#导入 NumPy 模块
import numpy as np
import ssl
#导入 pandas 模块
import pandas as pd
#导入图形化工具
import matplotlib.pyplot as plt
ssl._create_default_https_context = ssl._create_unverified_context
#下载后缀为 csv 的数据集,并将其作为训练集和测试集
```

```
df_train = pd.read_csv('https://storage.googleapis.com/tf-datasets/titanic/train.csv')
df_test = pd.read_csv('https://storage.googleapis.com/tf-datasets/titanic/eval.csv')
#分离标注字段
y_train = df_train.pop('survived')
y_test = df_test.pop('survived')
#显示年龄和存活率的关系
def calc_age_section(n, lim):
    return'[%.f,%.f)' % (lim*(n//lim), lim*(n//lim)+lim)   # map function
addone = pd.Series([calc_age_section(s, 10) for s in df_train.age])
df_train['ages'] = addone
pd.concat([df_train,y_train],
axis=1).groupby('ages').survived.mean().plot(kind='barh').set_xlabel('% survive');
plt.show()
```

(22)运行代码,得到年龄和存活率的关系如图10-6所示。

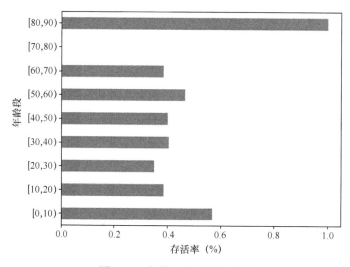

图10-6 年龄和存活率的关系

由图10-6可知,10岁以下的儿童和80岁以上的老人得到的生存机会较多。

(23)对典型数据进行单独分析和联合分析后,需要对数据进行预处理,本例中采用常见的one-hot编码进行处理并使用线性分类器,代码如下。

```
from __future__ import absolute_import, division, print_function
#导入NumPy模块
import numpy as np
import ssl
#导入pandas模块
import pandas as pd
#导入图形化工具
import matplotlib.pyplot as plt
#导入TensorFlow模块
import tensorflow as tf
ssl._create_default_https_context = ssl._create_unverified_context
```

第10章 TensorFlow 决策树

```python
#下载后缀为 csv 的数据集，并将其作为训练集和测试集
df_train = pd.read_csv('https://storage.googleapis.com/tf-datasets/titanic/train.csv')
df_test = pd.read_csv('https://storage.googleapis.com/tf-datasets/titanic/eval.csv')
#分离标注字段
y_train = df_train.pop('survived')
y_test = df_test.pop('survived')
#定义所需的数据列，分别定义分类型属性和数值型属性
CATEGORICAL_COLUMNS = ['sex', 'n_siblings_spouses', 'parch', 'class', 'deck',
                      'embark_town', 'alone']
NUMERIC_COLUMNS = ['age', 'fare']
#辅助函数，为给定数据列做 one-hot 编码
def one_hot_cat_column(feature_name, vocab):
    return tf.feature_column.indicator_column(
        tf.feature_column.categorical_column_with_vocabulary_list(feature_name,
                                                                  vocab))
#定义一个最后使用的列
feature_columns = []
for feature_name in CATEGORICAL_COLUMNS:
    #分类的属性都要做 one-hot 编码，然后加入数据列
    vocabulary = df_train[feature_name].unique()
    feature_columns.append(one_hot_cat_column(feature_name, vocabulary))
for feature_name in NUMERIC_COLUMNS:
    #数值类的属性直接入列
    feature_columns.append(tf.feature_column.numeric_column(feature_name,
                                                            dtype=tf.float32))
#将这个数据集作为一批载入
NUM_EXAMPLES = len(y_train)
#输入函数的构造函数
def make_input_fn(X, y, n_epochs=None, shuffle=True):
    def input_fn():
        dataset = tf.data.Dataset.from_tensor_slices((dict(X), y))
        #乱序
        if shuffle:
            dataset = dataset.shuffle(NUM_EXAMPLES)
        #训练时让数据重复尽量多的次数
        dataset = dataset.repeat(n_epochs)
        dataset = dataset.batch(NUM_EXAMPLES)
        return dataset
    return input_fn
#训练、评估所使用的数据输入函数，区别在于数据是否乱序及数据的迭代次数
train_input_fn = make_input_fn(df_train, y_train)
eval_input_fn = make_input_fn(df_test, y_test, shuffle=False, n_epochs=1)
#使用线性分类器
linear_est = tf.estimator.LinearClassifier(feature_columns)
#训练
linear_est.train(train_input_fn, max_steps=100)
```

```
#评估
result = linear_est.evaluate(eval_input_fn)
print( " -------------------------------- " )
print(pd.Series(result))
```

（24）代码的运行结果如下。

```
--------------------------------
accuracy                  0.765152
accuracy_baseline         0.625000
auc                       0.832844
auc_precision_recall      0.789631
average_loss              0.478908
label/mean                0.375000
loss                      0.478908
precision                 0.703297
prediction/mean           0.350790
recall                    0.646465
global_step             100.000000
```

说明：One-Hot 编码，又称为一位有效编码，可以将分类值映射到整数值并以二进制形式表示。

（25）引入 Boosted Trees 模型进行训练，使用 BoostedTreesClassifier 预测类的目标，代码如下。

```
from __future__ import absolute_import, division, print_function
#导入 NumPy 模块
import numpy as np
import ssl
#导入 pandas 模块
import pandas as pd
#导入图形化工具
import matplotlib.pyplot as plt
#导入 TensorFlow 模块
import tensorflow as tf
ssl._create_default_https_context = ssl._create_unverified_context
#下载后缀为 csv 的数据集，并将其作为训练集和测试集
df_train = pd.read_csv('https://storage.googleapis.com/tf-datasets/titanic/train.csv')
df_test = pd.read_csv('https://storage.googleapis.com/tf-datasets/titanic/eval.csv')
#分离标注字段
y_train = df_train.pop('survived')
y_test = df_test.pop('survived')
#定义所需的数据列，分别定义分类型属性和数值型属性
CATEGORICAL_COLUMNS = ['sex', 'n_siblings_spouses', 'parch', 'class', 'deck',
                       'embark_town', 'alone']
NUMERIC_COLUMNS = ['age', 'fare']
#辅助函数，为给定数据列做 one-hot 编码
```

```python
def one_hot_cat_column(feature_name, vocab):
    return tf.feature_column.indicator_column(
        tf.feature_column.categorical_column_with_vocabulary_list(feature_name,vocab))
#定义一个最后使用的列
feature_columns = []
for feature_name in CATEGORICAL_COLUMNS:
    #分类的属性都要做 one-hot 编码，然后加入数据列
    vocabulary = df_train[feature_name].unique()
    feature_columns.append(one_hot_cat_column(feature_name, vocabulary))
for feature_name in NUMERIC_COLUMNS:
    #数值类的属性直接入列
    feature_columns.append(tf.feature_column.numeric_column(feature_name,
                                                            dtype=tf.float32))
#将这个数据集作为一批载入
NUM_EXAMPLES = len(y_train)
#输入函数的构造函数
def make_input_fn(X, y, n_epochs=None, shuffle=True):
    def input_fn():
        dataset = tf.data.Dataset.from_tensor_slices((dict(X), y))
        #乱序
        if shuffle:
            dataset = dataset.shuffle(NUM_EXAMPLES)
        #训练时让数据重复尽量多的次数
        dataset = dataset.repeat(n_epochs)
        dataset = dataset.batch(NUM_EXAMPLES)
        return dataset
    return input_fn
#训练、评估所使用的数据输入函数，区别在于数据是否乱序及数据的迭代次数
train_input_fn = make_input_fn(df_train, y_train)
eval_input_fn = make_input_fn(df_test, y_test, shuffle=False, n_epochs=1)
#使用 Boosted Trees
n_batches = 1
est = tf.estimator.BoostedTreesClassifier(feature_columns,
                                          n_batches_per_layer=n_batches)
#训练
est.train(train_input_fn, max_steps=100)
#评估
result = est.evaluate(eval_input_fn)
print(" --------------------------------- ")
print(pd.Series(result))
```

（26）代码的运行结果如下。

```
---------------------------------
accuracy                    0.818182
```

```
accuracy_baseline        0.625000
auc                      0.859933
auc_precision_recall     0.850304
average_loss             0.419193
label/mean               0.375000
loss                     0.419193
precision                0.774194
prediction/mean          0.379677
recall                   0.727273
global_step            100.000000
dtype: float64
```

可以发现，在使用了 Boosted Trees 模型后，精确度提高。

（27）使用直方图对以上 2 种方法进行对比，代码如下。

```
from __future__ import absolute_import, division, print_function
#导入 NumPy 模块
import numpy as np
import ssl
#导入 pandas 模块
import pandas as pd
#导入图形化工具
import matplotlib.pyplot as plt
#导入 TensorFlow 模块
import tensorflow as tf
ssl._create_default_https_context = ssl._create_unverified_context
#下载后缀为 csv 的数据集，并将其作为训练集和测试集
df_train = pd.read_csv('https://storage.googleapis.com/tf-datasets/titanic/train.csv')
df_test = pd.read_csv('https://storage.googleapis.com/tf-datasets/titanic/eval.csv')
#分离标注字段
y_train = df_train.pop('survived')
y_test = df_test.pop('survived')
#定义所需的数据列，分别定义分类型属性和数值型属性
CATEGORICAL_COLUMNS = ['sex', 'n_siblings_spouses', 'parch', 'class', 'deck',
                       'embark_town', 'alone']
NUMERIC_COLUMNS = ['age', 'fare']
#辅助函数，为给定数据列做 one-hot 编码
def one_hot_cat_column(feature_name, vocab):
    return tf.feature_column.indicator_column(
        tf.feature_column.categorical_column_with_vocabulary_list(feature_name,
                                                                  vocab))
#定义一个最后使用的列
feature_columns = []
for feature_name in CATEGORICAL_COLUMNS:
    #分类的属性都要做 one-hot 编码，然后加入数据列
    vocabulary = df_train[feature_name].unique()
```

```python
            feature_columns.append(one_hot_cat_column(feature_name, vocabulary))
for feature_name in NUMERIC_COLUMNS:
    #数值类的属性直接入列
    feature_columns.append(tf.feature_column.numeric_column(feature_name,
                                                            dtype=tf.float32))
#将这个数据集作为一批载入
NUM_EXAMPLES = len(y_train)
#输入函数的构造函数
def make_input_fn(X, y, n_epochs=None, shuffle=True):
    def input_fn():
        dataset = tf.data.Dataset.from_tensor_slices((dict(X), y))
        #乱序
        if shuffle:
            dataset = dataset.shuffle(NUM_EXAMPLES)
        #训练时让数据重复尽量多的次数
        dataset = dataset.repeat(n_epochs)
        dataset = dataset.batch(NUM_EXAMPLES)
        return dataset
    return input_fn
#训练、评估所使用的数据输入函数，区别在于数据是否乱序及数据的迭代次数
train_input_fn = make_input_fn(df_train, y_train)
eval_input_fn = make_input_fn(df_test, y_test, shuffle=False, n_epochs=1)
#使用 Boosted Trees
n_batches = 1
est = tf.estimator.BoostedTreesClassifier(feature_columns,
                                          n_batches_per_layer=n_batches)
#训练
est.train(train_input_fn, max_steps=100)
#评估
result = est.evaluate(eval_input_fn)
#使用线性分类器
linear_est = tf.estimator.LinearClassifier(feature_columns)
#训练
linear_est.train(train_input_fn, max_steps=100)
#绘制直方图
pred_dicts1 = list(linear_est.predict(eval_input_fn))
pred_dicts2 = list(est.predict(eval_input_fn))
probs1 = pd.Series([pred['probabilities'][1] for pred in pred_dicts1])
probs2 = pd.Series([pred['probabilities'][1] for pred in pred_dicts2])
plt.figure(figsize=(14, 5))
plt.subplot(1, 2, 1)
probs1.plot(kind='hist', bins=20, title='Linear-est predicted probabilities');
plt.subplot(1, 2, 2)
probs2.plot(kind='hist', bins=20, title='Est predicted probabilities');
plt.show()
```

（28）运行代码，得到两种方法的对比如图10-7所示。

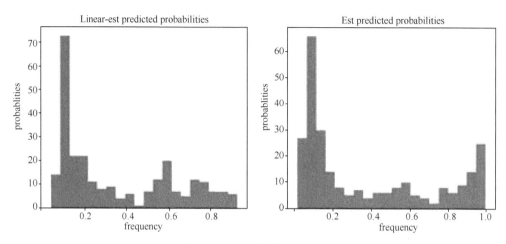

图10-7　两种方法的对比

（29）由于这里采用的是简单的数据集，所以用直方图可以显示清楚，但是随着分类算法复杂度的增加，一般要用接受者操作特性（Receiver Operating Characteristic）曲线来表示，简称 ROC 曲线。一般来说，一个分类器有 4 种分类情况，如表 10-2 所示。

表 10-2　分类情况

分类情况	说明	分类情况	说明
真阳性（TP）	判断为 1，实际也为 1	真阴性（TN）	判断为 0，实际也为 0
伪阳性（FP）	判断为 1，实际为 0	伪阴性（FN）	判断为 0，实际为 1

（30）ROC 曲线越接近图的左上角，表示与真阳性的极限值越接近，说明分类器的性能越好。绘制 ROC 曲线的代码如下。

```
from __future__ import absolute_import, division, print_function
#导入 NumPy 模块
import numpy as np
import ssl
#导入 pandas 模块
import pandas as pd
#导入图形化工具
import matplotlib.pyplot as plt
#导入 TensorFlow 模块
import tensorflow as tf
ssl._create_default_https_context = ssl._create_unverified_context
#下载后缀为 csv 的数据集，并将其作为训练集和测试集
df_train = pd.read_csv('https://storage.googleapis.com/tf-datasets/titanic/train.csv')
df_test = pd.read_csv('https://storage.googleapis.com/tf-datasets/titanic/eval.csv')
#分离标注字段
y_train = df_train.pop('survived')
y_test = df_test.pop('survived')
```

```python
#定义所需的数据列，分别定义分类型属性和数值型属性
CATEGORICAL_COLUMNS = ['sex', 'n_siblings_spouses', 'parch', 'class', 'deck',
                      'embark_town', 'alone']
NUMERIC_COLUMNS = ['age', 'fare']
#辅助函数，为给定数据列做 one-hot 编码
def one_hot_cat_column(feature_name, vocab):
    return tf.feature_column.indicator_column(
        tf.feature_column.categorical_column_with_vocabulary_list(feature_name,
                                                                  vocab))
#定义一个最后使用的列
feature_columns = []
for feature_name in CATEGORICAL_COLUMNS:
    #分类的属性都要做 one-hot 编码，然后加入数据列
    vocabulary = df_train[feature_name].unique()
    feature_columns.append(one_hot_cat_column(feature_name, vocabulary))
for feature_name in NUMERIC_COLUMNS:
    #数值类的属性直接入列
    feature_columns.append(tf.feature_column.numeric_column(feature_name,
                                                            dtype=tf.float32))
#将这个数据集作为一批载入
NUM_EXAMPLES = len(y_train)
#输入函数的构造函数
def make_input_fn(X, y, n_epochs=None, shuffle=True):
    def input_fn():
        dataset = tf.data.Dataset.from_tensor_slices((dict(X), y))
        #乱序
        if shuffle:
            dataset = dataset.shuffle(NUM_EXAMPLES)
        #训练时让数据重复尽量多的次数
        dataset = dataset.repeat(n_epochs)
        dataset = dataset.batch(NUM_EXAMPLES)
        return dataset
    return input_fn
#训练、评估所使用的数据输入函数，区别在于数据是否乱序及数据的迭代次数
train_input_fn = make_input_fn(df_train, y_train)
eval_input_fn = make_input_fn(df_test, y_test, shuffle=False, n_epochs=1)
#使用 Boosted Trees
n_batches = 1
est = tf.estimator.BoostedTreesClassifier(feature_columns,
                                          n_batches_per_layer=n_batches)
#训练
est.train(train_input_fn, max_steps=100)
#评估
result = est.evaluate(eval_input_fn)
#使用线性分类器
linear_est = tf.estimator.LinearClassifier(feature_columns)
```

```
#训练
linear_est.train(train_input_fn, max_steps=100)
pred_dicts1 = list(linear_est.predict(eval_input_fn))
pred_dicts2 = list(est.predict(eval_input_fn))
probs1 = pd.Series([pred['probabilities'][1] for pred in pred_dicts1])
probs2 = pd.Series([pred['probabilities'][1] for pred in pred_dicts2])
#绘制 ROC 曲线
from sklearn.metrics import roc_curve
#显示图像
def plot_roc(probs, title):
    fpr, tpr, _ = roc_curve(y_test, probs)
    plt.plot(fpr, tpr)
    plt.title(title)
    plt.xlabel('false positive rate')
    plt.ylabel('true positive rate')
    plt.xlim(0,)
    plt.ylim(0,)
plt.figure(figsize=(14, 5))
plt.subplot(1, 2, 1)
plot_roc(probs1, "Linear-est ROC ")
plt.subplot(1, 2, 2)
plot_roc(probs2, "Est ROC ")
plt.show()
```

（31）运行代码，得到 ROC 曲线的对比如图 10-8 所示。

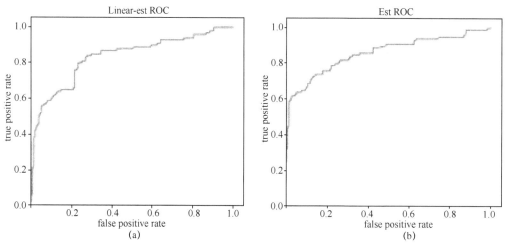

图 10-8　ROC 曲线的对比

从 ROC 曲线能够更直观地看出使用 Boosted Trees 的优势，即整体性能有很大的提高。决策树是数据挖掘和机器学习中常用的算法，灵活使用决策树能够提高学习的速度和精度。

第 11 章 TensorFlow 过拟合和欠拟合

过拟合（Overfitting）和欠拟合（Underfitting）是两种造成模型失真的常见情况，数据的大小及数据之间的关系"数据化"都有可能导致出现这两种情况。

11.1 过拟合和欠拟合的基本概念

过拟合：一个模型的学习能力"太强"，将使数据中的特点被模型过度捕获，导致模型的泛化能力下降。

欠拟合：数据集比较复杂，使模型的学习能力不足，不能总结出规律，导致模型的泛化能力弱。

以上两种会影响模型预测结果的情况的产生原因如下。

欠拟合：
（1）模型复杂度过低。
（2）特征量过少。

过拟合：
（1）建模样本选取有误。
（2）样本噪声干扰过大，导致机器将部分噪声误认为特征，扰乱了预设的分类规则。
（3）假设的模型无法合理存在或假设成立的条件实际并不成立。
（4）参数过多，模型复杂度过高。

11.2 过拟合和欠拟合

通过 11.1 节，读者应该已经对过拟合和欠拟合有了相应的了解，本节从实际例子出发，对过拟合和欠拟合的产生原因进行分析。本例将在电影评论分类网络上使用 Dense 图层作为基准来创建一个简单模型，并创建更小和更大的版本进行比较。

（1）准备并观察数据，代码如下。

```
from __future__ import absolute_import, division, print_function
import ssl
#导入图形化工具
import matplotlib.pyplot as plt
#以 TensorFlow 为基础构建 Keras
import tensorflow as tf
import tensorflow.keras as keras
#导入 NumPy 模块
```

```
import numpy as np
ssl._create_default_https_context = ssl._create_unverified_context
#定义词数目常量
NUM_WORDS = 10000
#导入数据集并以词数目常量划分训练集和测试集
(train_data, train_labels), (test_data, test_labels)=keras.datasets.imdb.load_data(num_words=NUM_WORDS)
#定义数据处理函数
def multi_hot_sequences(sequences, dimension):
    results = np.zeros((len(sequences), dimension))
    for i, word_indices in enumerate(sequences):
        results[i, word_indices] = 1.0
    return results
#对训练集和测试集进行处理
train_data = multi_hot_sequences(train_data, dimension=NUM_WORDS)
test_data = multi_hot_sequences(test_data, dimension=NUM_WORDS)
#显示训练集的第一个数据
plt.plot(train_data[0])
plt.show()
```

（2）运行代码，得到训练集的第一个数据如图 11-1 所示。

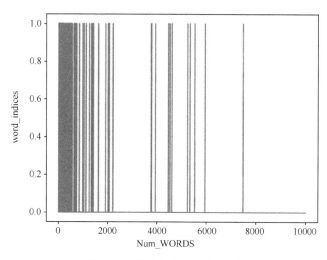

图 11-1　训练集的第一个数据

（3）构建一个基准模型，代码如下。

```
from __future__ import absolute_import, division, print_function
import ssl
#导入图形化工具
import matplotlib.pyplot as plt
#以 TensorFlow 为基础构建 Keras
import tensorflow as tf
import tensorflow.keras as keras
```

```python
#导入NumPy模块
import numpy as np
ssl._create_default_https_context = ssl._create_unverified_context
#定义词数目常量
NUM_WORDS = 10000
#导入数据集并以词数目常量划分训练集和测试集
(train_data, train_labels), (test_data, test_labels)=keras.datasets.imdb.load_data(num_words=NUM_WORDS)
#定义数据处理函数
def multi_hot_sequences(sequences, dimension):
    results = np.zeros((len(sequences), dimension))
    for i, word_indices in enumerate(sequences):
        results[i, word_indices] = 1.0
    return results
#对训练集和测试集进行处理
train_data = multi_hot_sequences(train_data, dimension=NUM_WORDS)
test_data = multi_hot_sequences(test_data, dimension=NUM_WORDS)
#使用时序模型构建基准模型
baseline_model = keras.Sequential(
[
    #全连接层
    layers.Dense(16, activation='relu', input_shape=(NUM_WORDS,)),
    layers.Dense(16, activation='relu'),
    layers.Dense(1, activation='sigmoid')
]
)
    #初始化基准模型参数
baseline_model.compile(optimizer='adam', loss='binary_crossentropy',
    metrics=['accuracy', 'binary_crossentropy'])
    #查看基准模型详情
baseline_model.summary()
```

（4）代码的运行结果如下。

```
Model: " sequential "
_____
Layer (type)                 Output Shape              Param #
=================================================================
dense (Dense)                (None, 16)                160016
_____
dense_1 (Dense)              (None, 16)                272
_____
dense_2 (Dense)              (None, 1)                 17
=================================================================
Total params: 160,305
```

```
Trainable params: 160,305
Non-trainable params: 0
_____
```

（5）对模型进行训练，本例中训练 20 次，代码如下。

```python
from __future__ import absolute_import, division, print_function
import ssl
#导入图形化工具
import matplotlib.pyplot as plt
#以 TensorFlow 为基础构建 Keras
import tensorflow as tf
import tensorflow.keras as keras
#导入 NumPy 模块
import numpy as np
ssl._create_default_https_context = ssl._create_unverified_context
#使用时序模型构建基准模型
baseline_model = keras.Sequential(
[
    #全连接层
    layers.Dense(16, activation='relu', input_shape=(NUM_WORDS,)),
    layers.Dense(16, activation='relu'),
    layers.Dense(1, activation='sigmoid')
]
)
#初始化基准模型参数
baseline_model.compile(optimizer='adam',
                    loss='binary_crossentropy',
                    metrics=['accuracy', 'binary_crossentropy'])
#制定基准模型训练计划并执行
baseline_history = baseline_model.fit(train_data, train_labels,
    epochs=20, batch_size=512, validation_data=(test_data, test_labels), verbose=2)
```

（6）代码的运行结果如下。

```
Train on 25000 samples, validate on 25000 samples
Epoch 1/20
5000/25000 - 13s - loss: 0.5010 - accuracy: 0.7927 - binary_crossentropy: 0.5010 - val_loss: 0.3537 - val_accuracy: 0.8717 - val_binary_crossentropy: 0.3537
...
Epoch 20/20
25000/25000 - 7s - loss: 0.0026 - accuracy: 1.0000 - binary_crossentropy: 0.0026 - val_loss: 0.8428 - val_accuracy: 0.8543 - val_binary_crossentropy: 0.8428
```

（7）使用相同的数据集构建一个小模型，代码如下。

```python
from __future__ import absolute_import, division, print_function
import ssl
#导入图形化工具
```

```python
import matplotlib.pyplot as plt
#以 TensorFlow 为基础构建 Keras
import tensorflow as tf
import tensorflow.keras as keras
#导入 NumPy 模块
import numpy as np
ssl._create_default_https_context = ssl._create_unverified_context
#定义词数目常量
NUM_WORDS = 10000
#导入数据集并以词数目常量划分训练集和测试集
(train_data, train_labels), (test_data, test_labels)=keras.datasets.imdb.load_data(num_words=NUM_WORDS)
#定义数据处理函数
def multi_hot_sequences(sequences, dimension):
    results = np.zeros((len(sequences), dimension))
    for i, word_indices in enumerate(sequences):
        results[i, word_indices] = 1.0
    return results
#对训练集和测试集进行处理
train_data = multi_hot_sequences(train_data, dimension=NUM_WORDS)
test_data = multi_hot_sequences(test_data, dimension=NUM_WORDS)
#使用时序模型构建基准模型
small_model = keras.Sequential(
[
 #全连接层
    layers.Dense(4, activation='relu', input_shape=(NUM_WORDS,)),
    layers.Dense(4, activation='relu'),
    layers.Dense(1, activation='sigmoid')
]
)
#初始化小模型参数
small_model.compile(optimizer='adam', loss='binary_crossentropy',
    metrics=['accuracy', 'binary_crossentropy'])
#输出小模型详情
small_model.summary()
```

(8) 代码的运行结果如下。

```
Model: "sequential"
_____
Layer (type)                 Output Shape              Param #
=================================================================
dense (Dense)                (None, 4)                 40004
_____
dense_1 (Dense)              (None, 4)                 20
```

```
dense_2 (Dense)                    (None, 1)                    5
=================================================================
Total params: 40,029
Trainable params: 40,029
Non-trainable params: 0
_____
```

(9)对模型进行训练,本例中训练20次,代码如下。

```
from __future__ import absolute_import, division, print_function
import ssl
#导入图形化工具
import matplotlib.pyplot as plt
#以 TensorFlow 为基础构建 Keras
import tensorflow as tf
import tensorflow.keras as keras
#导入 NumPy 模型
import numpy as np
ssl._create_default_https_context = ssl._create_unverified_context
#定义词数目常量
NUM_WORDS = 10000
#导入数据集并以词数目常量划分训练集和测试集
(train_data, train_labels), (test_data, test_labels)=keras.datasets.imdb.load_data(num_words=NUM_WORDS)
#定义数据处理函数
def multi_hot_sequences(sequences, dimension):
    results = np.zeros((len(sequences), dimension))
    for i, word_indices in enumerate(sequences):
        results[i, word_indices] = 1.0
    return results
#对训练集和测试集进行处理
train_data = multi_hot_sequences(train_data, dimension=NUM_WORDS)
test_data = multi_hot_sequences(test_data, dimension=NUM_WORDS)
#使用时序模型构建基准模型
small_model = keras.Sequential(
[
 #全连接层
    layers.Dense(4, activation='relu', input_shape=(NUM_WORDS,)),
    layers.Dense(4, activation='relu'),
    layers.Dense(1, activation='sigmoid')
]
)
#初始化小模型参数
small_model.compile(optimizer='adam', loss='binary_crossentropy',
```

```
        metrics=['accuracy', 'binary_crossentropy'])
    #制订训练计划并执行
    small_history = small_model.fit(train_data, train_labels,
        epochs=20, batch_size=512,
        validation_data=(test_data, test_labels), verbose=2)
```

（10）代码的运行结果如下。

```
Train on 25000 samples, validate on 25000 samples
Epoch 1/20
5000/25000 - 10s - loss: 0.5079 - accuracy: 0.8015 - binary_crossentropy: 0.5079 - val_loss: 0.3843 - val_accuracy: 0.8632 - val_binary_crossentropy: 0.3843
...
Epoch 20/20
25000/25000 - 8s - loss: 0.0298 - accuracy: 0.9964 - binary_crossentropy: 0.0298 - val_loss: 0.6009 - val_accuracy: 0.8532 - val_binary_crossentropy: 0.6009
```

（11）使用相同的数据集构建一个大模型，代码如下。

```
from __future__ import absolute_import, division, print_function
import ssl
#导入图形化工具
import matplotlib.pyplot as plt
#以 TensorFlow 为基础构建 Keras
import tensorflow as tf
import tensorflow.keras as keras
#导入 NumPy 模块
import numpy as np
ssl._create_default_https_context = ssl._create_unverified_context
#定义词数目常量
NUM_WORDS = 10000
#导入数据集并以词数目常量划分训练集和测试集
(train_data, train_labels), (test_data, test_labels)=keras.datasets.imdb.load_data(num_words=NUM_WORDS)
#定义数据处理函数
def multi_hot_sequences(sequences, dimension):
    results = np.zeros((len(sequences), dimension))
    for i, word_indices in enumerate(sequences):
        results[i, word_indices] = 1.0
    return results
#对训练集和测试集进行处理
train_data = multi_hot_sequences(train_data, dimension=NUM_WORDS)
test_data = multi_hot_sequences(test_data, dimension=NUM_WORDS)
#使用时序模型构建基准模型
big_model = keras.Sequential(
    [
    #全连接层
```

```
        layers.Dense(512, activation='relu', input_shape=(NUM_WORDS,)),
        layers.Dense(512, activation='relu'),
        layers.Dense(1, activation='sigmoid')
    ]
)
#初始化大模型参数
big_model.compile(optimizer='adam', loss='binary_crossentropy',
    metrics=['accuracy', 'binary_crossentropy'])
#输出大模型详情
big_model.summary()
```

(12) 代码的运行结果如下。

```
Model: "sequential"
_____
Layer (type)                 Output Shape              Param #
=================================================================
dense (Dense)                (None, 512)               5120512
_____
dense_1 (Dense)              (None, 512)               262656
_____
dense_2 (Dense)              (None, 1)                 513
=================================================================
Total params: 5,383,681
Trainable params: 5,383,681
Non-trainable params: 0
_____
```

(13) 对模型进行训练，本例中训练 20 次，代码如下。

```
from __future__ import absolute_import, division, print_function
import ssl
#导入图形化工具
import matplotlib.pyplot as plt
#以 TensorFlow 为基础构建 Keras
import tensorflow as tf
import tensorflow.keras as keras
#导入 NumPy 模块
import numpy as np
ssl._create_default_https_context = ssl._create_unverified_context
#定义词数目常量
NUM_WORDS = 10000
#导入数据集并以词数目常量划分训练集和测试集
(train_data, train_labels), (test_data, test_labels)=keras.datasets.imdb.load_data(num_words=NUM_WORDS)
#定义数据处理函数
```

```python
def multi_hot_sequences(sequences, dimension):
    results = np.zeros((len(sequences), dimension))
    for i, word_indices in enumerate(sequences):
        results[i, word_indices] = 1.0
    return results
#对训练集和测试集进行处理
train_data = multi_hot_sequences(train_data, dimension=NUM_WORDS)
test_data = multi_hot_sequences(test_data, dimension=NUM_WORDS)
#使用时序模型构建基准模型
big_model = keras.Sequential(
[
#全连接层
    layers.Dense(512, activation='relu', input_shape=(NUM_WORDS,)),
    layers.Dense(512, activation='relu'),
    layers.Dense(1, activation='sigmoid')
]
)
#初始化大模型参数
big_model.compile(optimizer='adam',
                  loss='binary_crossentropy',
                  metrics=['accuracy', 'binary_crossentropy'])
#制订训练计划并执行
big_history = big_model.fit(train_data, train_labels, epochs=20, batch_size=512,
    validation_data=(test_data, test_labels), verbose=2)
```

(14)代码的运行结果如下。

```
Train on 25000 samples, validate on 25000 samples
Epoch 1/20
5000/25000 - 18s - loss: 0.3408 - accuracy: 0.8563 - binary_crossentropy: 0.3408 - val_loss: 0.2962 - val_accuracy: 0.8780 - val_binary_crossentropy: 0.2962
...
Epoch 20/20
25000/25000 - 17s - loss: 1.1888e-05 - accuracy: 1.0000 - binary_crossentropy: 1.1888e-05 - val_loss: 0.9637 - val_accuracy: 0.8710 - val_binary_crossentropy: 0.9637
```

(15)对3个模型进行比较,代码如下。

```
from __future__ import absolute_import, division, print_function
import ssl
#导入图形化工具
import matplotlib.pyplot as plt
#以 TensorFlow 为基础构建 Keras
import tensorflow as tf
import tensorflow.keras as keras
```

```python
#导入NumPy模块
import numpy as np
ssl._create_default_https_context = ssl._create_unverified_context
#定义词数目常量
NUM_WORDS = 10000
#导入数据集并以词数目常量划分训练集和测试集
(train_data, train_labels), (test_data, test_labels)=keras.datasets.imdb.load_data(num_words=NUM_WORDS)
#定义数据处理函数
def multi_hot_sequences(sequences, dimension):
    results = np.zeros((len(sequences), dimension))
    for i, word_indices in enumerate(sequences):
        results[i, word_indices] = 1.0
    return results
#对训练集和测试集进行处理
train_data = multi_hot_sequences(train_data, dimension=NUM_WORDS)
test_data = multi_hot_sequences(test_data, dimension=NUM_WORDS)
#使用时序模型构建基准模型
baseline_model = keras.Sequential(
[
#全连接层
    layers.Dense(16, activation='relu', input_shape=(NUM_WORDS,)),
    layers.Dense(16, activation='relu'),
    layers.Dense(1, activation='sigmoid')
]
)
#初始化基准模型参数
baseline_model.compile(optimizer='adam', loss='binary_crossentropy',
    metrics=['accuracy', 'binary_crossentropy'])
#制订训练计划并执行
baseline_history = baseline_model.fit(train_data, train_labels,
    epochs=20, batch_size=512,
    validation_data=(test_data, test_labels), verbose=2)
#使用时序模型构建小模型
small_model = keras.Sequential(
[
#全连接层
    layers.Dense(4, activation='relu', input_shape=(NUM_WORDS,)),
    layers.Dense(4, activation='relu'),
    layers.Dense(1, activation='sigmoid')
]
```

```python
)
#初始化小模型参数
small_model.compile(optimizer='adam', loss='binary_crossentropy',
    metrics=['accuracy', 'binary_crossentropy'])
#制订训练计划并执行
small_history = small_model.fit(train_data, train_labels,
    epochs=20, batch_size=512,
    validation_data=(test_data, test_labels), verbose=2)
#使用时序模型构建大模型
big_model = keras.Sequential(
[
#全连接层
    layers.Dense(512, activation='relu', input_shape=(NUM_WORDS,)),
    layers.Dense(512, activation='relu'),
    layers.Dense(1, activation='sigmoid')
]
)
#初始化大模型参数
big_model.compile(optimizer='adam', loss='binary_crossentropy',
    metrics=['accuracy', 'binary_crossentropy'])
#制订训练计划并执行
big_history = big_model.fit(train_data, train_labels, epochs=20, batch_size=512,
    validation_data=(test_data, test_labels), verbose=2)
#显示图像
def plot_history(histories, key='binary_crossentropy'):
    plt.figure(figsize=(16,10))
    for name, history in histories:
        val = plt.plot(history.epoch, history.history['val_'+key],
            '--', label=name.title()+' val')
        plt.plot(history.epoch, history.history[key], color=val[0].get_color(),
            label=name.title()+' train')
    plt.xlabel('epochs')
    plt.ylabel(key.replace('_',' ').title())
    plt.legend()
    plt.xlim([0,max(history.epoch)])
plot_history([('baseline', baseline_history),
    ('small', small_history), ('big', big_history)])
plt.show()
```

（16）运行代码，得到3个模型的对比如图11-2所示。

由图11-2可知，在某时刻，大模型迅速开始过拟合，且过拟合严重。网络容量越大，就能够越快地对训练数据进行建模，但也越有可能出现过拟合。

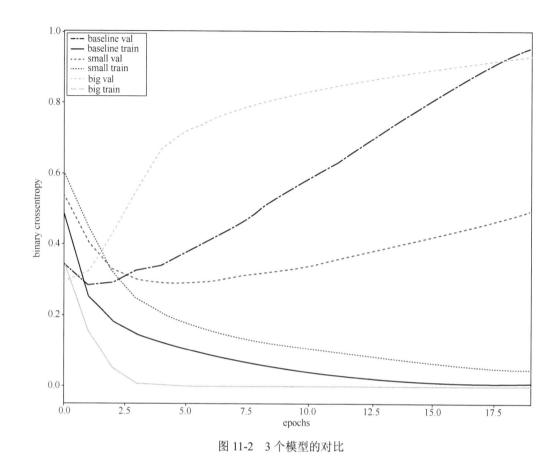

图 11-2　3 个模型的对比

11.3　优化方法

从 11.2 节可以看出，在没有优化模型的情况下，过拟合和欠拟合都会影响深度学习的预测结果，本节对如何优化模型进行探索。

11.3.1　dropout 优化方案

随机失活（dropout）是对具有深度结构的人工神经网络进行优化的方法。在学习过程中，其通过随机删除隐藏层的部分权重或输出来降低节点的相互依赖性，从而实现神经网络的正则化，降低其结构风险。

（1）构建 dropout 模型，并与 11.2 节的 3 个模型进行对比，代码如下。

```
from __future__ import absolute_import, division, print_function
import ssl
#导入图形化工具
import matplotlib.pyplot as plt
#以 TensorFlow 为基础构建 Keras
```

```python
import tensorflow as tf
import tensorflow.keras as keras
#导入NumPy模块
import numpy as np
ssl._create_default_https_context = ssl._create_unverified_context
#定义词数目常量
NUM_WORDS = 10000
#导入数据集并以词数目常量划分训练集和测试集
(train_data, train_labels), (test_data, test_labels)=keras.datasets.imdb.load_data(num_words=NUM_WORDS)
#定义数据处理函数
def multi_hot_sequences(sequences, dimension):
    results = np.zeros((len(sequences), dimension))
    for i, word_indices in enumerate(sequences):
        results[i, word_indices] = 1.0
    return results
#对训练集和测试集进行处理
train_data = multi_hot_sequences(train_data, dimension=NUM_WORDS)
test_data = multi_hot_sequences(test_data, dimension=NUM_WORDS)
#使用时序模型构建基准模型
baseline_model = keras.Sequential(
[
#全连接层
    layers.Dense(16, activation='relu', input_shape=(NUM_WORDS,)),
    layers.Dense(16, activation='relu'),
    layers.Dense(1, activation='sigmoid')
]
)
#初始化基准模型参数
baseline_model.compile(optimizer='adam', loss='binary_crossentropy',
    metrics=['accuracy', 'binary_crossentropy'])
#制订训练计划并执行
baseline_history = baseline_model.fit(train_data, train_labels,
    epochs=20, batch_size=512,
    validation_data=(test_data, test_labels), verbose=2)
#使用时序模型构建小模型
small_model = keras.Sequential(
[
#全连接层
    layers.Dense(4, activation='relu', input_shape=(NUM_WORDS,)),
    layers.Dense(4, activation='relu'),
```

```python
    layers.Dense(1, activation='sigmoid')
    ]
)
#初始化小模型参数
small_model.compile(optimizer='adam', loss='binary_crossentropy',
    metrics=['accuracy', 'binary_crossentropy'])
#制订训练计划并执行
small_history = small_model.fit(train_data, train_labels,
    epochs=20, batch_size=512,
    validation_data=(test_data, test_labels), verbose=2)
#使用时序模型构建大模型
big_model = keras.Sequential(
    [
    #全连接层
    layers.Dense(512, activation='relu', input_shape=(NUM_WORDS,)),
    layers.Dense(512, activation='relu'),
    layers.Dense(1, activation='sigmoid')
    ]
)
#初始化大模型参数
big_model.compile(optimizer='adam', loss='binary_crossentropy',
    metrics=['accuracy', 'binary_crossentropy'])
#制订训练计划并执行
big_history = big_model.fit(train_data, train_labels,
    epochs=20, batch_size=512,
    validation_data=(test_data, test_labels), verbose=2)
#dropout 优化模型
dpt_model = keras.Sequential(
    [
    layers.Dense(16, activation='relu', input_shape=(NUM_WORDS,)),
    #添加 dropout 层
    layers.Dropout(0.5),
    layers.Dense(16, activation='relu'),
    layers.Dropout(0.5),
    layers.Dense(1, activation='sigmoid')
    ]
)
#初始化 dropout 模型参数
dpt_model.compile(optimizer='adam', loss='binary_crossentropy',
    metrics=['accuracy', 'binary_crossentropy'])
#输出 dropout 模型详情
dpt_model.summary()
```

```
#对dropout模型进行训练
dpt_history = dpt_model.fit(train_data, train_labels,
    epochs=10, batch_size=512,
    validation_data=(test_data, test_labels), verbose=2)
#显示图像
def plot_history(histories, key='binary_crossentropy'):
    plt.figure(figsize=(16,10))
    for name, history in histories:
        val = plt.plot(history.epoch, history.history['val_'+key],
            '--', label=name.title()+' val')
        plt.plot(history.epoch, history.history[key], color=val[0].get_color(),
            label=name.title()+' train')
    plt.xlabel('epochs')
    plt.ylabel(key.replace('_',' ').title())
    plt.legend()
    plt.xlim([0,max(history.epoch)])
plot_history([('baseline', baseline_history), ('small', small_history),
    ('big', big_history), ('dropout',dpt_history)])
plt.show()
```

（2）代码的运行结果如下。

```
Model: "sequential_3"
_____
Layer (type)                 Output Shape              Param #
=================================================================
dense_9 (Dense)              (None, 16)                160016
_____
dropout (Dropout)            (None, 16)                0
_____
dense_10 (Dense)             (None, 16)                272
_____
dropout_1 (Dropout)          (None, 16)                0
_____
dense_11 (Dense)             (None, 1)                 17
=================================================================
Total params: 160,305
Trainable params: 160,305
Non-trainable params: 0
_____
```

（3）4个模型的对比如图11-3所示。

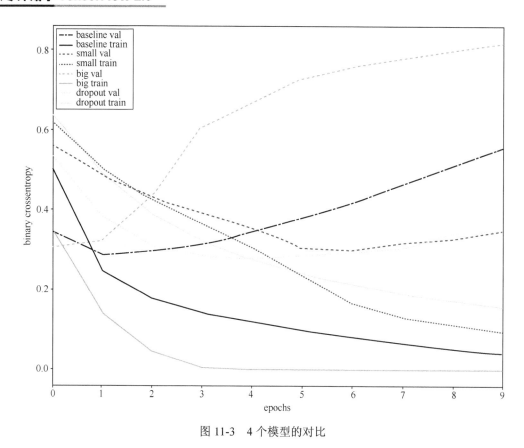

图 11-3 4 个模型的对比

从图 11-3 中可以看出，与前 3 个模型相比，dropout 模型的优化效果更好。

11.3.2　L2 正则化优化

在深度学习中，一般会在损失函数后面添加一个额外项（常用的是 L1-norm 和 L2-norm，即 L1 范数和 L2 范数）。添加这个额外项的目的是对损失函数进行惩罚，以调整复杂的模型。

在正则化中，L2 范数用于解决过拟合问题，可以避免模型过于复杂导致的过拟合问题。

（1）构造 L2 正则化模型，并与 11.3.1 节中的 4 个模型进行对比，代码如下。

```
from __future__ import absolute_import, division, print_function
import ssl
#导入图形化工具
import matplotlib.pyplot as plt
#以 TensorFlow 为基础构建 Keras
import tensorflow as tf
import tensorflow.keras as keras
#导入 NumPy 模块
import numpy as np
ssl._create_default_https_context = ssl._create_unverified_context
#定义词数目常量
```

```python
NUM_WORDS = 10000
#导入数据集并以词数目常量划分训练集和测试集
(train_data, train_labels), (test_data, test_labels)=keras.datasets.imdb.load_data(num_words=NUM_WORDS)
#定义数据处理函数
def multi_hot_sequences(sequences, dimension):
    results = np.zeros((len(sequences), dimension))
    for i, word_indices in enumerate(sequences):
        results[i, word_indices] = 1.0
    return results
#对训练集和测试集进行处理
train_data = multi_hot_sequences(train_data, dimension=NUM_WORDS)
test_data = multi_hot_sequences(test_data, dimension=NUM_WORDS)
#使用时序模型构建基准模型
baseline_model = keras.Sequential(
[
#全连接层
    layers.Dense(16, activation='relu', input_shape=(NUM_WORDS,)),
    layers.Dense(16, activation='relu'),
    layers.Dense(1, activation='sigmoid')
]
)
#初始化基准模型参数
baseline_model.compile(optimizer='adam', loss='binary_crossentropy',
    metrics=['accuracy', 'binary_crossentropy'])
#制订训练计划并执行
baseline_history = baseline_model.fit(train_data, train_labels,
    epochs=20, batch_size=512,
    validation_data=(test_data, test_labels), verbose=2)
#使用时序模型构建小模型
small_model = keras.Sequential(
[
#全连接层
    layers.Dense(4, activation='relu', input_shape=(NUM_WORDS,)),
    layers.Dense(4, activation='relu'),
    layers.Dense(1, activation='sigmoid')
]
)
#初始化小模型参数
small_model.compile(optimizer='adam', loss='binary_crossentropy',
    metrics=['accuracy', 'binary_crossentropy'])
#制订训练计划并执行
small_history = small_model.fit(train_data, train_labels,
```

```
        epochs=20, batch_size=512,
        validation_data=(test_data, test_labels), verbose=2)
#使用时序模型构建大模型
big_model = keras.Sequential(
[
#全连接层
    layers.Dense(512, activation='relu', input_shape=(NUM_WORDS,)),
    layers.Dense(512, activation='relu'),
    layers.Dense(1, activation='sigmoid')
]
)
#初始化大模型参数
big_model.compile(optimizer='adam',
    loss='binary_crossentropy', metrics=['accuracy', 'binary_crossentropy'])
#制订训练计划并执行
big_history = big_model.fit(train_data, train_labels, epochs=20, batch_size=512,
    validation_data=(test_data, test_labels), verbose=2)
#dropout 优化模型
dpt_model = keras.Sequential(
[
    layers.Dense(16, activation='relu', input_shape=(NUM_WORDS,)),
    #添加 dropout 层
    layers.Dropout(0.5),
    layers.Dense(16, activation='relu'),
    layers.Dropout(0.5),
    layers.Dense(1, activation='sigmoid')
]
)
#初始化 dropout 模型参数
dpt_model.compile(optimizer='adam', loss='binary_crossentropy',
    metrics=['accuracy', 'binary_crossentropy'])
#输出 dropout 模型详情
dpt_model.summary()
#对 dropout 模型进行训练
dpt_history = dpt_model.fit(train_data, train_labels, epochs=10, batch_size=512,
    validation_data=(test_data, test_labels), verbose=2)
#L2 正则化模型
l2_model = keras.Sequential(
[
#添加 L2 正则化模型的全连接层
    layers.Dense(16, kernel_regularizer=keras.regularizers.l2(0.001),
            activation='relu', input_shape=(NUM_WORDS,)),
    layers.Dense(16, kernel_regularizer=keras.regularizers.l2(0.001),
```

```
                    activation='relu'),
        layers.Dense(1, activation='sigmoid')
    ]
)
#初始化模型参数
l2_model.compile(optimizer='adam',
                 loss='binary_crossentropy',
                 metrics=['accuracy', 'binary_crossentropy'])
#输出模型详情
l2_model.summary()
#对模型进行训练
l2_history = l2_model.fit(train_data, train_labels, epochs=10, batch_size=512,
    validation_data=(test_data, test_labels), verbose=2)
#显示图像
def plot_history(histories, key='binary_crossentropy'):
    plt.figure(figsize=(16,10))
    for name, history in histories:
        val = plt.plot(history.epoch, history.history['val_'+key],
               '--', label=name.title()+' val')
        plt.plot(history.epoch, history.history[key], color=val[0].get_color(),
               label=name.title()+' train')
    plt.xlabel('epochs')
    plt.ylabel(key.replace('_',' ').title())
    plt.legend()
    plt.xlim([0,max(history.epoch)])
plot_history([('baseline', baseline_history), ('small', small_history),
    ('big', big_history), ('dropout',dpt_history), ('l2', l2_history)])
plt.show()
```

（2）运行代码，得到 L2 正则化模型的详情如下。

```
Model: " sequential_4 "
_____
Layer (type)                 Output Shape              Param #
=================================================================
dense_12 (Dense)             (None, 16)                160016
_____
dense_13 (Dense)             (None, 16)                272
_____
dense_14 (Dense)             (None, 1)                 17
=================================================================
Total params: 160,305
Trainable params: 160,305
Non-trainable params: 0
```

（3）5 个模型的对比如图 11-4 所示。

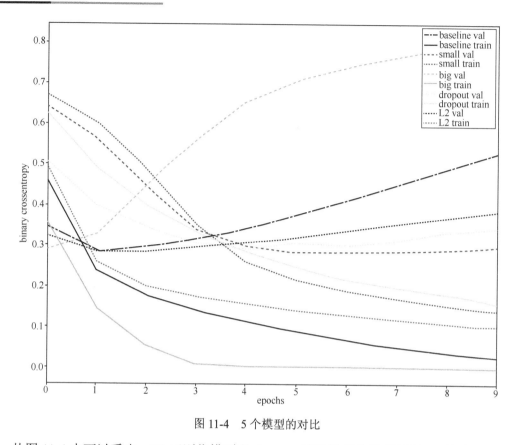

图 11-4 5 个模型的对比

从图 11-4 中可以看出，L2 正则化模型和 dropout 模型的优化效果较好。要根据不同场景的实际情况选择不同的优化方案。防止神经网络过拟合的常用方法有获取更多训练数据、减少网络容量、进行 dropout 优化和正则化优化等。

第12章 TensorFlow 结构化数据

在前面的章节中,已经使用 TensorFlow 进行了结构化操作。在实际构造机器学习模型之前,一般都会进行结构化操作。本章重点介绍如何结构化数据。

本章使用克利夫兰临床基金会提供的一个小数据集,其 CSV 中有几百行,每行描述一个患者,每列描述一个属性。可以使用此信息来预测患者是否患有心脏病,属于二元分类任务。

(1)准备数据集并对其进行验证,代码如下。

```
from __future__ import absolute_import, division, print_function
import ssl
#以 TensorFlow 为基础构建 Keras
import tensorflow as tf
import tensorflow.keras as keras
#导入 NumPy 模块
import numpy as np
#导入 pandas 模块
import pandas as pd
ssl._create_default_https_context = ssl._create_unverified_context
#使用 pandas 读取数据
URL = 'https://storage.googleapis.com/applied-dl/heart.csv'
dataframe = pd.read_csv(URL)
#显示前 5 行数据
head = dataframe.head()
print (head)
```

(2)代码的运行结果如下。

	age	sex	cp	trestbps	chol	fbs	restecg	thalach	exang	oldpeak	slope	ca	thal	target
0	63	1	1	145	233	1	2	150	0	2.3	3	0	fixed	0
1	67	1	4	160	286	0	2	108	1	1.5	2	3	normal	1
2	67	1	4	120	229	0	2	129	1	2.6	2	2	reversible	0
3	37	1	3	130	250	0	0	187	0	3.5	3	0	normal	0
4	41	0	2	130	204	0	2	172	0	1.4	1	0	normal	0

(3)划分训练集和测试集,代码如下。

```
from __future__ import absolute_import, division, print_function
import ssl
#以 TensorFlow 为基础构建 Keras
import tensorflow as tf
import tensorflow.keras as keras
#导入 NumPy 模块
import numpy as np
```

```
#导入pandas模块
import pandas as pd
ssl._create_default_https_context = ssl._create_unverified_context
#使用pandas读取数据
URL = 'https://storage.googleapis.com/applied-dl/heart.csv'
dataframe = pd.read_csv(URL)
#划分训练集和测试集
train, test = train_test_split(dataframe, test_size=0.2)
train, val = train_test_split(train, test_size=0.2)
print(len(train), 'train examples')
print(len(val), 'validation examples')
print(len(test), 'test examples')
```

（4）代码的运行结果如下。

```
193 train examples
49 validation examples
61 test examples
```

（5）构建pipeline，代码如下。

```
from __future__ import absolute_import, division, print_function
import ssl
#以TensorFlow为基础构建Keras
import tensorflow as tf
import tensorflow.keras as keras
#导入NumPy模块
import numpy as np
#导入pandas模块
import pandas as pd
from sklearn.model_selection import train_test_split
ssl._create_default_https_context = ssl._create_unverified_context
#使用pandas读取数据
URL = 'https://storage.googleapis.com/applied-dl/heart.csv'
dataframe = pd.read_csv(URL)
#划分训练集和测试集
train, test = train_test_split(dataframe, test_size=0.2)
train, val = train_test_split(train, test_size=0.2)
#定义使数据集随机分布的函数
def df_to_dataset(dataframe, shuffle=True, batch_size=32):
    dataframe = dataframe.copy()
    labels = dataframe.pop('target')
    ds = tf.data.Dataset.from_tensor_slices((dict(dataframe), labels))
    if shuffle:
        ds = ds.shuffle(buffer_size=len(dataframe))
    ds = ds.batch(batch_size)
    return ds
#定义batch_size常量
```

```
batch_size = 5
#使用定义的函数对数据进行处理
train_ds = df_to_dataset(train, batch_size=batch_size)
val_ds = df_to_dataset(val, shuffle=False, batch_size=batch_size)
test_ds = df_to_dataset(test, shuffle=False, batch_size=batch_size)
for feature_batch, label_batch in train_ds.take(1):
    print('Every feature:', list(feature_batch.keys()))
    print('A batch of ages:', feature_batch['age'])
    print('A batch of targets:', label_batch )
```

(6) 代码的运行结果如下。

```
Every feature: ['age', 'sex', 'cp', 'trestbps', 'chol', 'fbs', 'restecg', 'thalach', 'exang', 'oldpeak', 'slope', 'ca', 'thal']
A batch of ages: tf.Tensor([52 50 48 59 63], shape=(5,), dtype=int32)
A batch of targets: tf.Tensor([0 0 1 1 1], shape=(5,), dtype=int32)
```

12.1 数字列

特征列的输出是模型的输入。数字列是最简单的列，它用于表示真正有价值的特征。在使用此列时，模型将从数据框中接收未更改的列值。

(1) 本节使用下面的代码对数字列进行介绍。

```
from __future__ import absolute_import, division, print_function
import ssl
#以 TensorFlow 为基础构建 Keras
import tensorflow as tf
import tensorflow.keras as keras
#导入 NumPy 模块
import numpy as np
#导入 pandas 模块
import pandas as pd
from sklearn.model_selection import train_test_split
from tensorflow import feature_column
ssl._create_default_https_context = ssl._create_unverified_context
#使用 pandas 读取数据
URL = 'https://storage.googleapis.com/applied-dl/heart.csv'
dataframe = pd.read_csv(URL)
#划分训练集和测试集
train, test = train_test_split(dataframe, test_size=0.2)
train, val = train_test_split(train, test_size=0.2)
#定义使数据集随机分布的函数
def df_to_dataset(dataframe, shuffle=True, batch_size=32):
    dataframe = dataframe.copy()
    labels = dataframe.pop('target')
```

```
            ds = tf.data.Dataset.from_tensor_slices((dict(dataframe), labels))
        if shuffle:
            ds = ds.shuffle(buffer_size=len(dataframe))
        ds = ds.batch(batch_size)
        return ds
    #定义batch_size常量
    batch_size = 5
    #使用定义的函数对数据进行处理
    train_ds = df_to_dataset(train, batch_size=batch_size)
    val_ds = df_to_dataset(val, shuffle=False, batch_size=batch_size)
    test_ds = df_to_dataset(test, shuffle=False, batch_size=batch_size)
    example_batch = next(iter(train_ds))[0]
    #定义数据输出函数
    def print_data(feature_column):
        feature_layer = layers.DenseFeatures(feature_column)
        print (feature_layer(example_batch).numpy())
        #构建数字列
    age = feature_column.numeric_column("age")
    print_data(age)
```

(2)代码的运行结果如下。

```
[[57.]
 [42.]
 [54.]
 [58.]
 [39.]]
```

12.2 bucketized 列

在希望将数字直接输入模型的情况下，需要根据数值范围将数字的值分成不同的类别。这时可以使用 bucketized 列将年龄分成几个桶。

(1)本节使用下面的代码对 bucketized 列进行介绍。

```
from __future__ import absolute_import, division, print_function
import ssl
#以 TensorFlow 为基础构建 Keras
import tensorflow as tf
import tensorflow.keras as keras
#导入 NumPy 模块
import numpy as np
#导入 pandas 模块
import pandas as pd
from sklearn.model_selection import train_test_split
```

```python
from tensorflow import feature_column
ssl._create_default_https_context = ssl._create_unverified_context
#使用 pandas 读取数据
URL = 'https://storage.googleapis.com/applied-dl/heart.csv'
dataframe = pd.read_csv(URL)
#划分训练集和测试集
train, test = train_test_split(dataframe, test_size=0.2)
train, val = train_test_split(train, test_size=0.2)
#定义使数据集随机分布的函数
def df_to_dataset(dataframe, shuffle=True, batch_size=32):
    dataframe = dataframe.copy()
    labels = dataframe.pop('target')
    ds = tf.data.Dataset.from_tensor_slices((dict(dataframe), labels))
    if shuffle:
        ds = ds.shuffle(buffer_size=len(dataframe))
    ds = ds.batch(batch_size)
    return ds
#定义 batch_size 常量
batch_size = 5
#使用定义的函数对数据进行处理
train_ds = df_to_dataset(train, batch_size=batch_size)
val_ds = df_to_dataset(val, shuffle=False, batch_size=batch_size)
test_ds = df_to_dataset(test, shuffle=False, batch_size=batch_size)
example_batch = next(iter(train_ds))[0]
age = feature_column.numeric_column("age")
#定义数据输出函数
def print_data(feature_column):
    feature_layer = layers.DenseFeatures(feature_column)
    print (feature_layer(example_batch).numpy())
#构建 bucketized 列
age_buckets = feature_column.bucketized_column(age, boundaries=[
    18, 25, 30, 35, 40, 50
])
print_data(age_buckets)
```

（2）代码的运行结果如下。

```
[[0. 0. 0. 0. 0. 1. 0.]
 [0. 0. 0. 0. 0. 0. 1.]
 [0. 0. 0. 0. 0. 0. 1.]
 [0. 0. 0. 0. 0. 0. 1.]
 [0. 0. 0. 0. 0. 1. 0.]]
```

12.3 类别列

在数据集中，thal 表示字符串，需要直接将字符串提供给模型。在这种情况下，必须先将数据映射到数值。类别列提供了一种将字符串表示为单热矢量的方法。

（1）本节使用下面的代码对类别列进行介绍。

```
from __future__ import absolute_import, division, print_function
import ssl
#以 TensorFlow 为基础构建 Keras
import tensorflow as tf
import tensorflow.keras as keras
#导入 NumPy 模块
import numpy as np
#导入 pandas 模块
import pandas as pd
from sklearn.model_selection import train_test_split
from tensorflow import feature_column
ssl._create_default_https_context = ssl._create_unverified_context
#使用 pandas 读取数据
URL = 'https://storage.googleapis.com/applied-dl/heart.csv'
dataframe = pd.read_csv(URL)
#划分训练集和测试集
train, test = train_test_split(dataframe, test_size=0.2)
train, val = train_test_split(train, test_size=0.2)
#定义使数据集随机分布的函数
def df_to_dataset(dataframe, shuffle=True, batch_size=32):
    dataframe = dataframe.copy()
    labels = dataframe.pop('target')
    ds = tf.data.Dataset.from_tensor_slices((dict(dataframe), labels))
    if shuffle:
        ds = ds.shuffle(buffer_size=len(dataframe))
    ds = ds.batch(batch_size)
    return ds
#定义 batch_size 常量
batch_size = 5
#使用定义的函数对数据进行处理
train_ds = df_to_dataset(train, batch_size=batch_size)
val_ds = df_to_dataset(val, shuffle=False, batch_size=batch_size)
test_ds = df_to_dataset(test, shuffle=False, batch_size=batch_size)
example_batch = next(iter(train_ds))[0]
age = feature_column.numeric_column("age")
#定义数据输出函数
```

```python
def print_data(feature_column):
    feature_layer = layers.DenseFeatures(feature_column)
    print (feature_layer(example_batch).numpy())
#构建类别列
thal = feature_column.categorical_column_with_vocabulary_list('thal', ['fixed', 'normal', 'reversible'])
thal_one_hot = feature_column.indicator_column(thal)
print_data(thal_one_hot)
```

(2)代码的运行结果如下。

```
[[0. 1. 0.]
 [0. 1. 0.]
 [0. 0. 1.]
 [0. 0. 1.]
 [0. 0. 1.]]
```

12.4 嵌入列

嵌入列将数据表示为低维密集向量。其中，每个单元格可以包含任意数字，而不仅是 0 或 1。嵌入列数据可以在多类别的情况下训练神经网络。

(1)本节使用下面的代码对嵌入列进行介绍。

```python
from __future__ import absolute_import, division, print_function
import ssl
#以 TensorFlow 为基础构建 Keras
import tensorflow as tf
import tensorflow.keras as keras
#导入 NumPy 模块
import numpy as np
#导入 pandas 模块
import pandas as pd
from sklearn.model_selection import train_test_split
from tensorflow import feature_column
ssl._create_default_https_context = ssl._create_unverified_context
#使用 pandas 读取数据
URL = 'https://storage.googleapis.com/applied-dl/heart.csv'
dataframe = pd.read_csv(URL)
#划分训练集和测试集
train, test = train_test_split(dataframe, test_size=0.2)
train, val = train_test_split(train, test_size=0.2)
#定义使数据集随机分布的函数
def df_to_dataset(dataframe, shuffle=True, batch_size=32):
    dataframe = dataframe.copy()
    labels = dataframe.pop('target')
```

```
            ds = tf.data.Dataset.from_tensor_slices((dict(dataframe), labels))
            if shuffle:
                ds = ds.shuffle(buffer_size=len(dataframe))
            ds = ds.batch(batch_size)
            return ds
        #定义batch_size常量
        batch_size = 5
        #使用定义的函数对数据进行处理
        train_ds = df_to_dataset(train, batch_size=batch_size)
        val_ds = df_to_dataset(val, shuffle=False, batch_size=batch_size)
        test_ds = df_to_dataset(test, shuffle=False, batch_size=batch_size)
        example_batch = next(iter(train_ds))[0]
        age = feature_column.numeric_column("age")
        #定义数据输出函数
        def print_data(feature_column):
            feature_layer = layers.DenseFeatures(feature_column)
            print (feature_layer(example_batch).numpy())
        #构建嵌入列
        thal = feature_column.categorical_column_with_vocabulary_list('thal', ['fixed', 'normal', 'reversible'])
        thal_embedding = feature_column.embedding_column(thal, dimension=8)
        print_data(thal_embedding)
```

（2）代码的运行结果如下。

```
[[ 0.18184556   0.04679189   0.27516183   0.10210931  -0.3724993    0.19421828
  -0.21873671   0.27726513]
 [ 0.18184556   0.04679189   0.27516183   0.10210931  -0.3724993    0.19421828
  -0.21873671   0.27726513]
 [-0.12618273  -0.5095625   -0.15360346  -0.30314672  -0.07281923   0.4768423
  -0.08894793  -0.02658685]
 [ 0.18184556   0.04679189   0.27516183   0.10210931  -0.3724993    0.19421828
  -0.21873671   0.27726513]
 [ 0.18184556   0.04679189   0.27516183   0.10210931  -0.3724993    0.19421828
  -0.21873671   0.27726513]]
```

从结果可以看出，当分类列具有许多可能的值时，最好使用嵌入列对数据进行处理。

12.5 哈希特征列

可以使用 categorical_column_with_hash_bucket 表示具有大量值的分类列。哈希特征列计算输入的哈希值，并选择 hash_bucket_size 存储桶来编码字符串。它可以在不依赖词汇表的情况下减小实际类别的数量。

（1）本节使用下面的代码对哈希特征列进行介绍。

```
from __future__ import absolute_import, division, print_function
```

第12章 TensorFlow 结构化数据

```python
import ssl
#以 TensorFlow 为基础构建 Keras
import tensorflow as tf
import tensorflow.keras as keras
#导入 NumPy 模块
import numpy as np
#导入 pandas 模块
import pandas as pd
from sklearn.model_selection import train_test_split
from tensorflow import feature_column
ssl._create_default_https_context = ssl._create_unverified_context
#使用 pandas 读取数据
URL = 'https://storage.googleapis.com/applied-dl/heart.csv'
dataframe = pd.read_csv(URL)
#划分训练集和测试集
train, test = train_test_split(dataframe, test_size=0.2)
train, val = train_test_split(train, test_size=0.2)
#定义使数据集随机分布的函数
def df_to_dataset(dataframe, shuffle=True, batch_size=32):
    dataframe = dataframe.copy()
    labels = dataframe.pop('target')
    ds = tf.data.Dataset.from_tensor_slices((dict(dataframe), labels))
    if shuffle:
        ds = ds.shuffle(buffer_size=len(dataframe))
    ds = ds.batch(batch_size)
    return ds
#定义 batch_size 常量
batch_size = 5
#使用定义的函数对数据进行处理
train_ds = df_to_dataset(train, batch_size=batch_size)
val_ds = df_to_dataset(val, shuffle=False, batch_size=batch_size)
test_ds = df_to_dataset(test, shuffle=False, batch_size=batch_size)
example_batch = next(iter(train_ds))[0]
age = feature_column.numeric_column("age")
#定义数据输出函数
def print_data(feature_column):
    feature_layer = layers.DenseFeatures(feature_column)
    print (feature_layer(example_batch).numpy())
#构建哈希特征列
thal = feature_column.categorical_column_with_vocabulary_list('thal', ['fixed', 'normal', 'reversible'])
thal_hashed=feature_column.categorical_column_with_hash_bucket('thal',hash_bucket_size=1000)
print_data(feature_column.indicator_column(thal_hashed))
```

（2）代码的运行结果如下。

```
[[0. 0. 0. ... 0. 0. 0.]
 [0. 0. 0. ... 0. 0. 0.]
 [0. 0. 0. ... 0. 0. 0.]
 [0. 0. 0. ... 0. 0. 0.]
 [0. 0. 0. ... 0. 0. 0.]]
```

说明：该方法的缺点是数据之间可能会存在冲突，不同的字符串可能被映射到同一个桶上。因此，在使用的过程中需要对数据进行校验。

12.6 交叉功能列

交叉功能列使模型能够为每个特征组合学习权重。本例使用 age_buckets 和 thal 进行验证。

（1）本节使用下面的代码对交叉功能列进行介绍。

```
from __future__ import absolute_import, division, print_function
import ssl
#以 TensorFlow 为基础构建 Keras
import tensorflow as tf
import tensorflow.keras as keras
#导入 NumPy 模块
import numpy as np
#导入 pandas 模块
import pandas as pd
from sklearn.model_selection import train_test_split
from tensorflow import feature_column
ssl._create_default_https_context = ssl._create_unverified_context
#使用 pandas 读取数据
URL = 'https://storage.googleapis.com/applied-dl/heart.csv'
dataframe = pd.read_csv(URL)
#划分训练集和测试集
train, test = train_test_split(dataframe, test_size=0.2)
train, val = train_test_split(train, test_size=0.2)
#定义使数据集随机分布的函数
def df_to_dataset(dataframe, shuffle=True, batch_size=32):
    dataframe = dataframe.copy()
    labels = dataframe.pop('target')
    ds = tf.data.Dataset.from_tensor_slices((dict(dataframe), labels))
    if shuffle:
        ds = ds.shuffle(buffer_size=len(dataframe))
    ds = ds.batch(batch_size)
    return ds
```

```
#定义batch_size常量
batch_size = 5
#使用定义的函数对数据进行处理
train_ds = df_to_dataset(train, batch_size=batch_size)
val_ds = df_to_dataset(val, shuffle=False, batch_size=batch_size)
test_ds = df_to_dataset(test, shuffle=False, batch_size=batch_size)
example_batch = next(iter(train_ds))[0]
age = feature_column.numeric_column("age")
#定义数据输出函数
def print_data(feature_column):
    feature_layer = layers.DenseFeatures(feature_column)
    print (feature_layer(example_batch).numpy())
#构建交叉功能列
thal = feature_column.categorical_column_with_vocabulary_list('thal', ['fixed', 'normal', 'reversible'])
age_buckets = feature_column.bucketized_column(age, boundaries=[
    18, 25, 30, 35, 40, 50
])
crossed_feature = feature_column.crossed_column([age_buckets, thal], hash_bucket_size=1000)
print_data(feature_column.indicator_column(crossed_feature))
```

（2）代码的运行结果如下。

```
[[0. 0. 0. ... 0. 0. 0.]
 [0. 0. 0. ... 0. 0. 0.]
 [0. 0. 0. ... 0. 0. 0.]
 [0. 0. 0. ... 0. 0. 0.]
 [0. 0. 0. ... 0. 0. 0.]]
```

说明：crossed_column 不会构建所有可能组合的完整表（可能非常大）。在 hashed_column 的支持下，这类表的大小是可供选择的。

12.7 结构化数据的使用

本节对前面几节中介绍的结构化数据示例进行总结并举例进行说明，以加深读者对结构化数据的理解。

（1）对示例数据集进行结构化处理，并构建相应的模型进行训练，本例的训练次数为 5 次，代码如下。

```
from __future__ import absolute_import, division, print_function
import ssl
#以 TensorFlow 为基础构建 Keras
import tensorflow as tf
import tensorflow.keras as keras
```

```python
#导入 NumPy 模块
import numpy as np
#导入 pandas 模块
import pandas as pd
from sklearn.model_selection import train_test_split
from tensorflow import feature_column
ssl._create_default_https_context = ssl._create_unverified_context
#使用 pandas 读取数据
URL = 'https://storage.googleapis.com/applied-dl/heart.csv'
dataframe = pd.read_csv(URL)
#划分训练集和测试集
train, test = train_test_split(dataframe, test_size=0.2)
train, val = train_test_split(train, test_size=0.2)
#定义使数据集随机分布的函数
def df_to_dataset(dataframe, shuffle=True, batch_size=32):
    dataframe = dataframe.copy()
    labels = dataframe.pop('target')
    ds = tf.data.Dataset.from_tensor_slices((dict(dataframe), labels))
    if shuffle:
        ds = ds.shuffle(buffer_size=len(dataframe))
    ds = ds.batch(batch_size)
    return ds
#定义 batch_size 常量
batch_size = 5
#使用定义的函数对数据进行处理
train_ds = df_to_dataset(train, batch_size=batch_size)
val_ds = df_to_dataset(val, shuffle=False, batch_size=batch_size)
test_ds = df_to_dataset(test, shuffle=False, batch_size=batch_size)
example_batch = next(iter(train_ds))[0]
age = feature_column.numeric_column("age")
#定义数据输出函数
def print_data(feature_column):
    feature_layer = layers.DenseFeatures(feature_column)
    print (feature_layer(example_batch).numpy())
#初始化一个 feature_columns
feature_columns = []
#构建数字列
for header in ['age', 'trestbps', 'chol', 'thalach', 'oldpeak', 'slope', 'ca']:
    feature_columns.append(feature_column.numeric_column(header))
#构建 bucketized 列
age_buckets = feature_column.bucketized_column(age, boundaries=[18, 25, 30, 35, 40, 45, 50, 55, 60, 65])
feature_columns.append(age_buckets)
```

```python
#构建类别列
thal = feature_column.categorical_column_with_vocabulary_list(
    'thal', ['fixed', 'normal', 'reversible'])
thal_one_hot = feature_column.indicator_column(thal)
feature_columns.append(thal_one_hot)
#构建嵌入列
thal_embedding = feature_column.embedding_column(thal, dimension=8)
feature_columns.append(thal_embedding)
#构建哈希特征列
crossed_feature = feature_column.crossed_column([age_buckets, thal], hash_bucket_size=1000)
feature_columns.append(thal_hashed)
#构建交叉功能列
crossed_feature = feature_column.indicator_column(crossed_feature)
feature_columns.append(crossed_feature)
#构建特征层
feature_layer = tf.keras.layers.DenseFeatures(feature_columns)
batch_size = 32
train_ds = df_to_dataset(train, batch_size=batch_size)
val_ds = df_to_dataset(val, shuffle=False, batch_size=batch_size)
test_ds = df_to_dataset(test, shuffle=False, batch_size=batch_size)
#构建模型并训练
model = tf.keras.Sequential([
    feature_layer,
    layers.Dense(128, activation='relu'),
    layers.Dense(128, activation='relu'),
    layers.Dense(1, activation='sigmoid')
])
#初始化模型参数
model.compile(optimizer='adam', loss='binary_crossentropy', metrics=['accuracy'])
#对模型进行训练
model.fit(train_ds, validation_data=val_ds, epochs=5)
```

（2）代码的运行结果如下。

```
Epoch 1/5
        7/Unknown - 2s 224ms/step - loss: 1.1059 - accuracy: 0.6477
7/7 [==============================] - 2s 275ms/step - loss: 1.1059 - accuracy: 0.6477 - val_loss: 0.0000e+00 - val_accuracy: 0.0000e+00
...
Epoch 5/5
1/7 [===>..........................] - ETA: 0s - loss: 0.5578 - accuracy: 0.7812
7/7 [==============================] - 0s 8ms/step - loss: 0.5160 - accuracy: 0.7668 - val_loss: 0.5898 - val_accuracy: 0.7347
```

（3）训练结束后对模型进行评估并输出结果，代码如下。

```python
from __future__ import absolute_import, division, print_function
import ssl
#以 TensorFlow 为基础构建 Keras
import tensorflow as tf
import tensorflow.keras as keras
#导入 NumPy 模块
import numpy as np
#导入 pandas 模块
import pandas as pd
from sklearn.model_selection import train_test_split
from tensorflow import feature_column
ssl._create_default_https_context = ssl._create_unverified_context
#使用 pandas 读取数据
URL = 'https://storage.googleapis.com/applied-dl/heart.csv'
dataframe = pd.read_csv(URL)
#划分训练集和测试集
train, test = train_test_split(dataframe, test_size=0.2)
train, val = train_test_split(train, test_size=0.2)
#定义使数据集随机分布的函数
def df_to_dataset(dataframe, shuffle=True, batch_size=32):
    dataframe = dataframe.copy()
    labels = dataframe.pop('target')
    ds = tf.data.Dataset.from_tensor_slices((dict(dataframe), labels))
    if shuffle:
        ds = ds.shuffle(buffer_size=len(dataframe))
    ds = ds.batch(batch_size)
    return ds
#定义 batch_size 常量
batch_size = 5
#使用定义的函数对数据进行处理
train_ds = df_to_dataset(train, batch_size=batch_size)
val_ds = df_to_dataset(val, shuffle=False, batch_size=batch_size)
test_ds = df_to_dataset(test, shuffle=False, batch_size=batch_size)
example_batch = next(iter(train_ds))[0]
age = feature_column.numeric_column("age")
#定义数据输出函数
def print_data(feature_column):
    feature_layer = layers.DenseFeatures(feature_column)
    print (feature_layer(example_batch).numpy())
#初始化一个 feature_columns
feature_columns = []
#构建数字列
```

```python
for header in ['age', 'trestbps', 'chol', 'thalach', 'oldpeak', 'slope', 'ca']:
    feature_columns.append(feature_column.numeric_column(header))
#构建bucketized列
age_buckets = feature_column.bucketized_column(age, boundaries=[18, 25, 30, 35, 40, 45, 50, 55, 60, 65])
feature_columns.append(age_buckets)
#构建类别列
thal = feature_column.categorical_column_with_vocabulary_list(
    'thal', ['fixed', 'normal', 'reversible'])
thal_one_hot = feature_column.indicator_column(thal)
feature_columns.append(thal_one_hot)
#构建嵌入列
thal_embedding = feature_column.embedding_column(thal, dimension=8)
feature_columns.append(thal_embedding)
#构建哈希特征列
crossed_feature = feature_column.crossed_column([age_buckets, thal], hash_bucket_size=1000)
feature_columns.append(thal_hashed)
#构建交叉功能列
crossed_feature = feature_column.indicator_column(crossed_feature)
feature_columns.append(crossed_feature)
#构建特征层
feature_layer = tf.keras.layers.DenseFeatures(feature_columns)
batch_size = 32
train_ds = df_to_dataset(train, batch_size=batch_size)
val_ds = df_to_dataset(val, shuffle=False, batch_size=batch_size)
test_ds = df_to_dataset(test, shuffle=False, batch_size=batch_size)
#构建模型并训练
model = tf.keras.Sequential([feature_layer,
    layers.Dense(128, activation='relu'),
    layers.Dense(128, activation='relu'),
    layers.Dense(1, activation='sigmoid')
])
#初始化模型参数
model.compile(optimizer='adam', loss='binary_crossentropy', metrics=['accuracy'])
#对模型进行训练
model.fit(train_ds, validation_data=val_ds, epochs=5)
#对模型进行评估,并输出结果
loss, accuracy = model.evaluate(test_ds)
print("Accuracy", accuracy)
print("Loss", loss)
```

(4)代码的运行结果如下。

```
Epoch 1/5
    7/Unknown - 1s 204ms/step - loss: 3.5172 - accuracy: 0.5544
7/7 [==============================] - 2s 253ms/step - loss: 3.5172 - accuracy: 0.5544 - val_loss: 0.0000e+00 - val_accuracy: 0.0000e+00
...
Epoch 5/5
1/7 [===>..........................] - ETA: 0s - loss: 0.5771 - accuracy: 0.6562
7/7 [==============================] - 0s 8ms/step - loss: 0.5824 - accuracy: 0.7150 - val_loss: 0.6013 - val_accuracy: 0.6735
2/2 [==============================] - 0s 4ms/step - loss: 0.6318 - accuracy: 0.6557
Accuracy 0.6557377
Loss 0.6318106949329376
```

本节的示例对本章介绍的结构化数据进行了综合介绍,希望能够加深读者对 TensorFlow 2.0 结构化数据的理解。

第 13 章 TensorFlow 回归

回归是在建模过程中通过不断迭代数据来发现规律,并根据该规律对新数据进行预测的一种方法。在实际项目中,回归是一种重要的数据解析方法,本章对这种方法进行介绍。

13.1 一元线性回归

线性回归是一种利用数理统计中的回归分析来确定两种或两种以上变量间相互依赖的定量关系的统计分析方法,其运用十分广泛,表达形式为 $y = w'x + e$,e 服从均值为 0 的正态分布。

若回归中只有一个自变量和一个因变量,且两者的关系可以用一条直线近似表示,则这种回归为一元线性回归。

(1) 本节构建一个一元线性回归模型,准备数据集并对其进行验证,代码如下。

```
import keras
import numpy as np
import matplotlib.pyplot as plt
#按顺序构建 Sequential 模型
from keras.models import Sequential
#Sequential 是模型结构,包含输入层、隐藏层、输出层
from keras.optimizers import SGD
import pandas as pd
import numpy as np
#返回指定间隔的数字
X = np.linspace(-1, 1, 300)
#根据公式构造数据点,并加入一些噪声点
Y = 0.8 * X + 20 + np.random.normal(0, 0.05, (300,))
#显示数据点图
plt.scatter(X, Y)
plt.show()
```

(2) 运行代码,得到数据点图如图 13-1 所示。

(3) 由于是一元线性回归模型,所以构建一个一层的模型,代码如下。

```
import keras
import numpy as np
import matplotlib.pyplot as plt
#按顺序构建 Sequential 模型
from keras.models import Sequential
#Sequential 是模型结构,包含输入层、隐藏层、输出层
from keras.optimizers import SGD
```

```python
import pandas as pd
from keras.layers import Dense
import numpy as np
#返回指定间隔的数字
X = np.linspace(-1, 1, 300)
#根据公式构造数据点,并加入一些噪声点
Y = 0.8 * X + 20 + np.random.normal(0, 0.05, (300,))
#取前160个点,并将其作为训练集
X_train, Y_train = X[:160], Y[:160]
#取后140个点,并将其作为测试集
X_test, Y_test = X[160:], Y[160:]
#构建一个一层的模型
model = Sequential()
model.add(Dense(output_dim=1, input_dim=1))
#选择误差计算方法和优化器
model.compile(loss='mse', optimizer='sgd')
#显示模型详情
model.summary()
```

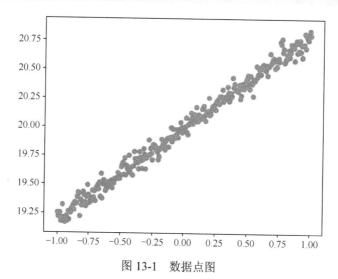

图 13-1 数据点图

(4) 代码的运行结果如下。

```
Model: "sequential_1"
_____
Layer (type)                 Output Shape              Param #
=================================================================
dense_1 (Dense)              (None, 1)                 2
=================================================================
Total params: 2
Trainable params: 2
Non-trainable params: 0
_____
```

(5) 执行训练计划,本例中训练100次,代码如下。

```python
import keras
import numpy as np
import matplotlib.pyplot as plt
#按顺序构建Sequential模型
from keras.models import Sequential
#Sequential是模型结构,包含输入层、隐藏层、输出层
from keras.optimizers import SGD
import pandas as pd
from keras.layers import Dense
import numpy as np
#返回指定间隔的数字
X = np.linspace(-1, 1, 300)
#根据公式构造数据点,并加入一些噪声点
Y = 0.8 * X + 20 + np.random.normal(0, 0.05, (300,))
#取前160个点,并将其作为训练集
X_train, Y_train = X[:160], Y[:160]
#取后140个点,并将其作为测试集
X_test, Y_test = X[160:], Y[160:]
#构建一个一层的模型
model = Sequential()
model.add(Dense(output_dim=1, input_dim=1))
#选择误差计算方法和优化器
model.compile(loss='mse', optimizer='sgd')
#制订训练计划,训练100次,令每次训练的batch为10并随机打乱数据
history = model.fit(X, Y, verbose=1, epochs=100,batch_size=10, shuffle=True,
    validation_data=(X_test,Y_test))
```

(6) 代码的运行结果如下。

```
Train on 300 samples, validate on 140 samples
Epoch 1/100
300/300 [==============================] - 0s 332us/step - loss: 236.8547 - val_loss: 129.7934
...
Epoch 100/100
300/300 [==============================] - 0s 124us/step - loss: 0.0027 - val_loss: 0.00
```

(7) 预测训练结果并显示预测图,代码如下。

```python
import keras
import numpy as np
import matplotlib.pyplot as plt
#按顺序构建Sequential模型
from keras.models import Sequential
#Sequential是模型结构,包含输入层、隐藏层、输出层
from keras.optimizers import SGD
```

```python
import pandas as pd
from keras.layers import Dense
import numpy as np
#返回指定间隔的数字
X = np.linspace(-1, 1, 300)
#根据公式构造数据点,并加入一些噪声点
Y = 0.8 * X + 20 + np.random.normal(0, 0.05, (300,))
#取前160个点,并将其作为训练集
X_train, Y_train = X[:160], Y[:160]
#取后140个点,并将其作为测试集
X_test, Y_test = X[160:], Y[160:]
#构建一个一层的模型
model = Sequential()
model.add(Dense(output_dim=1, input_dim=1))
#选择误差计算方法和优化器
model.compile(loss='mse', optimizer='sgd')
#制订训练计划,训练100次,令每次训练的batch为10并随机打乱数据
history = model.fit(X, Y, verbose=1, epochs=100,batch_size=10, shuffle=True,
    validation_data=(X_test,Y_test))
#显示预测图
Y_pred = model.predict(X_test,batch_size=1)
plt.scatter(X_test, Y_test)
plt.plot(X_test, Y_pred,'r.')
plt.show()
```

(8)运行代码,得到预测图如图13-2所示。

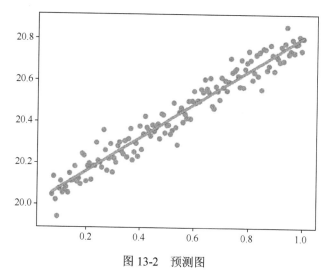

图13-2 预测图

从图13-2中可以看出,在受噪声干扰的情况下,该一元线性回归模型可以训练出与实际数据相似的数据。

13.2 多元线性回归

在 13.1 节中,使用 TensorFlow 2.0 构建了简单的一元线性回归模型。本节使用 TensorFlow 2.0 对存在 n 个($n \geq 2$)自变量的多元线性回归问题进行分析。

(1)本节构建一个多元线性回归模型,准备数据集并对其进行验证,代码如下。

```
import keras
import numpy as np
import matplotlib.pyplot as plt
#按顺序构建 Sequential 模型
from keras.models import Sequential
#Sequential 是模型结构,包含输入层、隐含层、输出层
from keras.optimizers import SGD
import pandas as pd
from keras.layers import Dense
import numpy as np
#构造多参数曲线,并加入噪声点
X = np.linspace(-2 * np.pi, 2 * np.pi, 300)
X = np.reshape(X, [X.__len__(), 1])
noise = np.random.rand(X.__len__(), 1) * 0.1
Y = np.sin(X) + noise
#显示多元数据集图像
plt.scatter(X, Y)
plt.show()
```

(2)运行代码,得到多元数据集如图 13-3 所示。

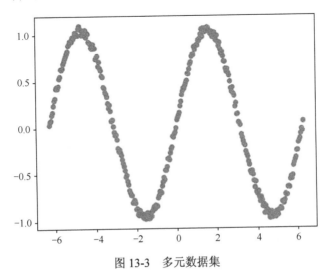

图 13-3 多元数据集

(3)构建多层时序模型,代码如下。

```
import keras
```

```python
import numpy as np
import matplotlib.pyplot as plt
#按顺序构建 Sequential 模型
from keras.models import Sequential
#Sequential 是模型结构,包含输入层、隐藏层、输出层
from keras.optimizers import SGD
import pandas as pd
from keras.layers import Dense
import numpy as np
#构造多参数曲线,并加入噪声点
X = np.linspace(-2 * np.pi,2 * np.pi, 300)
X = np.reshape(X, [X.__len__(), 1])
noise = np.random.rand(X.__len__(), 1) * 0.1
Y = np.sin(X) + noise
#取前 160 个点,并将其作为训练集
X_train, Y_train = X[:160], Y[:160]
#取后 140 个点,并将其作为测试集
X_test, Y_test = X[160:], Y[160:]
#构建一个多层时序模型
models = Sequential()
models.add(Dense(100, init='uniform',activation='relu' ,input_dim=1))
models.add(Dense(50, activation='relu'))
models.add(Dense(1,activation='tanh'))
models.compile(optimizer='rmsprop', loss='mse',metrics=[ " accuracy " ] )
#显示模型详情
models.summary()
```

(4)代码的运行结果如下。

```
Model: " sequential_1 "
_____
Layer (type)                 Output Shape              Param #
=================================================================
dense_1 (Dense)              (None, 100)               200
_____
dense_2 (Dense)              (None, 50)                5050
_____
dense_3 (Dense)              (None, 1)                 51
=================================================================
Total params: 5,301
Trainable params: 5,301
Non-trainable params: 0
_____
```

(5)执行训练计划,本例中训练 100 次,代码如下。

```
import keras
```

```python
import numpy as np
import matplotlib.pyplot as plt
#按顺序构建 Sequential 模型
from keras.models import Sequential
#Sequential 是模型结构，包含输入层、隐藏层、输出层
from keras.optimizers import SGD
import pandas as pd
from keras.layers import Dense
import numpy as np
#构造多参数曲线，并加入噪声点
X = np.linspace(-2 * np.pi,2 * np.pi, 300)
X = np.reshape(X, [X.__len__(), 1])
noise = np.random.rand(X.__len__(), 1) * 0.1
Y = np.sin(X) + noise
#取前 160 个点，并将其作为训练集
X_train, Y_train = X[:160], Y[:160]
#取后 140 个点，并将其作为测试集
X_test, Y_test = X[160:], Y[160:]
#构建一个多层时序模型
models = Sequential()
models.add(Dense(100, init='uniform',activation='relu' ,input_dim=1))
models.add(Dense(50, activation='relu'))
models.add(Dense(1,activation='tanh'))
models.compile(optimizer='rmsprop', loss='mse',metrics=[ " accuracy " ] )
#制订训练计划，训练 100 次，令每次训练的 batch 为 10 并随机打乱数据
history = models.fit(X, Y, verbose=1, epochs=100,batch_size=10, shuffle=True,
    validation_data=(X_test,Y_test))
```

（6）代码的运行结果如下。

```
Train on 300 samples, validate on 140 samples
Epoch 1/100
300/300 [==============================] - 0s 1ms/step - loss: 0.4130 - accuracy: 0.0000e+00 - val_loss: 0.4071 - val_accuracy: 0.0000e+00

...
Epoch 100/100
300/300 [==============================] - 0s 161us/step - loss: 0.0447 - accuracy: 0.0000e+00 - val_loss: 0.0304 - val_accuracy: 0.0000e+00
```

（7）预测训练结果并显示多元线性回归预测图，代码如下。

```python
import keras
import numpy as np
import matplotlib.pyplot as plt
#按顺序构建 Sequential 模型
from keras.models import Sequential
#Sequential 是模型结构，包含输入层、隐藏层、输出层
from keras.optimizers import SGD
```

```
import pandas as pd
from keras.layers import Dense
import numpy as np
#构造多参数曲线,并加入噪声点
X = np.linspace(-2 * np.pi,2 * np.pi, 300)
X = np.reshape(X, [X.__len__(), 1])
noise = np.random.rand(X.__len__(), 1) * 0.1
Y = np.sin(X) + noise
#取前160个点,并将其作为训练集
X_train, Y_train = X[:160], Y[:160]
#取后140个点,并将其作为测试集
X_test, Y_test = X[160:], Y[160:]
#构建一个多层时序模型
models = Sequential()
models.add(Dense(100, init='uniform',activation='relu' ,input_dim=1))
models.add(Dense(50, activation='relu'))
models.add(Dense(1,activation='tanh'))
models.compile(optimizer='rmsprop', loss='mse',metrics=[ " accuracy " ] )
#制订训练计划,训练100次,令每次训练的batch为10并随机打乱数据
history = models.fit(X, Y, verbose=1, epochs=100,batch_size=10, shuffle=True,
    validation_data=(X_test,Y_test))
#显示多元线性回归预测图
Y_pred = models.predict(X_test,batch_size=1)
plt.scatter(X_test, Y_test)
plt.plot(X_test, Y_pred,'r.')
plt.show()
```

(8)运行代码,得到多元线性回归预测图如图13-4所示。

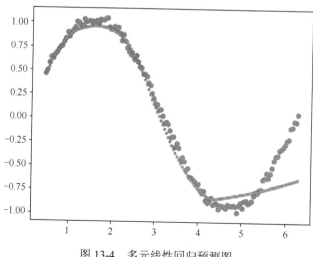

图13-4 多元线性回归预测图

从图 13-4 中可以看出，在一定阶段内，预测数据和实际数据相似，但噪声的干扰使数据出现了偏离。

13.3 汽车油耗回归示例

前面已经对回归模型进行了简单的介绍，本节对一个复杂的数据集进行回归分析。

本例中采用的数据集是 Auto MPG 数据集，本节根据该数据集对 20 世纪 70 年代末和 80 年代初的汽车燃油效率进行预测。

（1）下载数据集，代码如下。

```
from tensorflow import keras
import pandas as pd
import ssl
ssl._create_default_https_context = ssl._create_unverified_context
#定义下载地址
dataset_path=keras.utils.get_file("auto-mpg.data","http://archive.ics.uci.edu/ml/machine-learning-databases/auto-mpg/auto-mpg.data")
#规定列的名称
column_names = ['MPG','Cylinders','Displacement','Horsepower','Weight',
    'Acceleration', 'Model Year', 'Origin']
#读取数据
raw_dataset = pd.read_csv(dataset_path, names=column_names,
    na_values = "?", comment='\t', sep=" ", skipinitialspace=True)
#读取后 5 行数据并输出
tail = raw_dataset.tail()
print(tail)
```

（2）代码的运行结果如下。

```
     MPG   Cylinders  Displacement  Horsepower  Weight  Acceleration  Model Year  Origin
393  27.0      4         140.0        86.0      2790.0      15.6          82          1
394  44.0      4          97.0        52.0      2130.0      24.6          82          2
395  32.0      4         135.0        84.0      2295.0      11.6          82          1
396  28.0      4         120.0        79.0      2625.0      18.6          82          1
397  31.0      4         119.0        82.0      2720.0      19.4          82          1
```

（3）上述结果是一个数据集列表，其释义如表 13-1 所示。

表 13-1 释义表

ID	数据编号	ID	数据编号
MPG	每加仑汽油能行驶的英里数	Weight	汽车质量
Cylinders	汽车气缸数	Acceleration	汽车加速度
Displacement	汽车排量	Model Year	上市年份
Horsepower	汽车马力	Origin	产地

（4）读取全部数据，代码如下。

```
from tensorflow import keras
import pandas as pd
import ssl
ssl._create_default_https_context = ssl._create_unverified_context
#定义下载地址
dataset_path=keras.utils.get_file("auto-mpg.data","http://archive.ics.uci.edu/ml/machine-learning-databases/auto-mpg/auto-mpg.data")
#规定列的名称
column_names = ['MPG','Cylinders','Displacement','Horsepower','Weight',
    'Acceleration', 'Model Year', 'Origin']
#读取数据
raw_dataset = pd.read_csv(dataset_path, names=column_names,
    na_values = "?", comment='\t', sep=" ", skipinitialspace=True)
#读取全部数据
All = raw_dataset
print(All)
```

（5）代码的运行结果如下。

	MPG	Cylinders	Displacement	Horsepower	Weight	Acceleration	Model Year	Origin
0	18.0	8	307.0	130.0	3504.0	2.0	70	1
1	15.0	8	350.0	165.0	3693.0	11.5	70	1
2	18.0	8	318.0	150.0	3436.0	11.0	70	1
3	16.0	8	304.0	150.0	3433.0	12.0	70	1
4	17.0	8	302.0	140.0	3449.0	10.5	70	1
...	...	...	...	...	...	...	...	...
393	27.0	4	140.0	86.0	2790.0	15.6	82	1
394	44.0	4	97.0	52.0	2130.0	24.6	82	2
395	32.0	4	135.0	84.0	2295.0	11.6	82	1
396	28.0	4	120.0	79.0	2625.0	18.6	82	1
397	31.0	4	119.0	82.0	2720.0	19.4	82	1

[398 rows x 8 columns]

（6）实际上，在本例使用的数据集中有无效数据，而本例通过在使用pandas读取数据时对na_values进行相应的"?"设置，避免了无效数据的出现。可以使用下面的代码对数据集中是否有无效数据进行校验。

```
from tensorflow import keras
import pandas as pd
import ssl
ssl._create_default_https_context = ssl._create_unverified_context
#定义下载地址
dataset_path=keras.utils.get_file("auto-mpg.data","http://archive.ics.uci.edu/ml/machine-learning-databases/auto-mpg/auto-mpg.data")
#规定列的名称
column_names = ['MPG','Cylinders','Displacement','Horsepower','Weight',
    'Acceleration', 'Model Year', 'Origin']
#读取数据
```

```
raw_dataset = pd.read_csv(dataset_path, names=column_names,
    na_values = "?", comment='\t', sep=" ", skipinitialspace=True)
#显示无效数据
Nan = raw_dataset.isna().sum()
print(Nan)
```

（7）代码的运行结果如下，可以看出，在 Horsepower 字段中有 6 个无效数据。

```
MPG             0
Cylinders       0
Displacement    0
Horsepower      6
Weight          0
Acceleration    0
Model Year      0
Origin          0
dtype: int64
```

（8）从数据集中去除无效的数据，并处理数据分类问题，代码如下。

```
from tensorflow import keras
import pandas as pd
import ssl
ssl._create_default_https_context = ssl._create_unverified_context
#定义下载地址
dataset_path=keras.utils.get_file("auto-mpg.data","http://archive.ics.uci.edu/ml/machine-learning-databases/auto-mpg/auto-mpg.data")
#规定列的名称
column_names = ['MPG','Cylinders','Displacement','Horsepower','Weight',
    'Acceleration', 'Model Year', 'Origin']
#读取数据
raw_dataset = pd.read_csv(dataset_path, names=column_names,
    na_values = "?", comment='\t', sep=" ", skipinitialspace=True)
#复制一份数据做后续操作
dataset = raw_dataset.copy()
#去掉无效数据
dataset = dataset.dropna()
#对 Origin 数据做 one-hot 编码，将数字转换成相应的产地
origin = dataset.pop('Origin')
dataset['USA'] = (origin == 1)*1.0
dataset['Europe'] = (origin == 2)*1.0
dataset['Japan'] = (origin == 3)*1.0
#显示处理后的数据
print(dataset)
```

（9）代码的运行结果如下。

	MPG	Cylinders	Displacement	Horsepower	Weight	Acceleration	Model Year	USA	Europe	Japan
0	18.0	8	307.0	130.0	3504.0	12.0	70	1.0	0.0	0.0

1	15.0	8	350.0	165.0	3693.0	11.5	70	1.0	0.0	0.0
2	18.0	8	318.0	150.0	3436.0	11.0	70	1.0	0.0	0.0
3	16.0	8	304.0	150.0	3433.0	12.0	70	1.0	0.0	0.0
4	17.0	8	302.0	140.0	3449.0	10.5	70	1.0	0.0	0.0
...	...	...	...	...	...	...	...	...	...	...
393	27.0	4	140.0	86.0	2790.0	15.6	82	1.0	0.0	0.0
394	44.0	4	97.0	52.0	2130.0	24.6	82	0.0	1.0	0.0
395	32.0	4	135.0	84.0	2295.0	11.6	82	1.0	0.0	0.0
396	28.0	4	120.0	79.0	2625.0	18.6	82	1.0	0.0	0.0
397	31.0	4	119.0	82.0	2720.0	19.4	82	1.0	0.0	0.0

（10）在处理完数据后，需要将数据集划分为训练集和测试集，代码如下。

```
from tensorflow import keras
import pandas as pd
import ssl
ssl._create_default_https_context = ssl._create_unverified_context
#定义下载地址
dataset_path=keras.utils.get_file("auto-mpg.data","http://archive.ics.uci.edu/ml/machine-learning-databases/auto-mpg/auto-mpg.data")
#规定列的名称
column_names = ['MPG','Cylinders','Displacement','Horsepower','Weight',
    'Acceleration', 'Model Year', 'Origin']
#读取数据
raw_dataset = pd.read_csv(dataset_path, names=column_names,
    na_values = "?", comment='\t', sep=" ", skipinitialspace=True)
#复制一份数据做后续操作
dataset = raw_dataset.copy()
#去掉无效数据
dataset = dataset.dropna()
#对Origin数据做one-hot编码，将数字转换成相应的产地
origin = dataset.pop('Origin')
dataset['USA'] = (origin == 1)*1.0
dataset['Europe'] = (origin == 2)*1.0
dataset['Japan'] = (origin == 3)*1.0
#通过frac参数随机分配80%的数据，并将其作为训练集
#random_state是随机数的种子，使用固定的值保证每次构建数据集的数据一致
train_dataset = dataset.sample(frac=0.8, random_state=0)
#余下的20%为测试集
test_dataset = dataset.drop(train_dataset.index)
#获取训练集的统计信息
train_stats = train_dataset.describe()
print("train_stats")
print(train_stats)
print("\n")
```

```python
#获取测试集的统计信息
test_stats = test_dataset.describe()
print("test_stats")
print(test_stats)
```

(11) 代码的运行结果如下。

train_stats

	MPG	Cylinders	Displacement	Horsepower	Weight	Acceleration	Model Year	USA	Europe	Japan
count	314.000000	314.000000	314.000000	314.000000	314.000000	314.000000	314.000000	314.000000	314.000000	314.000000
mean	23.310510	5.477707	195.318471	104.869427	2990.251592	15.559236	75.898089	0.624204	0.178344	0.197452
std	7.728652	1.699788	104.331589	38.096214	843.898596	2.789230	3.675642	0.485101	0.383413	0.398712
min	10.000000	3.000000	68.000000	46.000000	1649.000000	8.000000	70.000000	0.000000	0.000000	0.000000
25%	17.000000	4.000000	105.500000	76.250000	2256.500000	13.800000	73.000000	0.000000	0.000000	0.000000
50%	22.000000	4.000000	151.000000	94.500000	2822.500000	15.500000	76.000000	1.000000	0.000000	0.000000
75%	28.950000	8.000000	265.750000	128.000000	3608.000000	17.200000	79.000000	1.000000	0.000000	0.000000
max	46.600000	8.000000	455.000000	225.000000	5140.000000	24.800000	82.000000	1.000000	1.000000	1.000000

test_stats

	MPG	Cylinders	Displacement	Horsepower	Weight	Acceleration	Model Year	USA	Europe	Japan
count	78.000000	78.000000	78.000000	78.000000	78.000000	78.000000	78.000000	78.000000	78.000000	78.000000
mean	23.991026	5.448718	190.762821	102.858974	2926.589744	15.469231	76.307692	0.628205	0.153846	0.217949
std	8.133563	1.740633	106.494733	40.255265	874.900416	2.649298	3.721847	0.486412	0.363137	0.415525
min	9.000000	3.000000	70.000000	48.000000	1613.000000	8.500000	70.000000	0.000000	0.000000	0.000000
25%	18.000000	4.000000	98.000000	74.250000	2159.500000	13.700000	73.000000	0.000000	0.000000	0.000000
50%	24.000000	4.000000	138.000000	90.000000	2692.500000	15.250000	76.000000	1.000000	0.000000	0.000000
75%	29.875000	8.000000	292.000000	121.750000	3706.500000	17.000000	79.750000	1.000000	0.000000	0.000000
max	44.300000	8.000000	400.000000	230.000000	4746.000000	22.200000				

```
    82.000000  1.000000     1.000000     1.000000
```
（12）根据要分析的情况对划分好的数据集进行整理，代码如下。

```
from tensorflow import keras
import pandas as pd
import ssl
ssl._create_default_https_context = ssl._create_unverified_context
#定义下载地址
dataset_path=keras.utils.get_file("auto-mpg.data","http://archive.ics.uci.edu/ml/machine-learning-databases/auto-mpg/auto-mpg.data")
#规定列的名称
column_names = ['MPG','Cylinders','Displacement','Horsepower','Weight',
    'Acceleration', 'Model Year', 'Origin']
#读取数据
raw_dataset = pd.read_csv(dataset_path, names=column_names,
    na_values = "?", comment='\t', sep=" ", skipinitialspace=True)
#复制一份数据做后续操作
dataset = raw_dataset.copy()
#去掉无效数据
dataset = dataset.dropna()
#对 Origin 数据做 one-hot 编码，将数字转换成相应的产地
origin = dataset.pop('Origin')
dataset['USA'] = (origin == 1)*1.0
dataset['Europe'] = (origin == 2)*1.0
dataset['Japan'] = (origin == 3)*1.0
#通过 frac 参数随机分配 80%的数据，并将其作为训练集
#random_state 是随机数的种子，使用固定的值保证每次构建数据集的数据一致
train_dataset = dataset.sample(frac=0.8, random_state=0)
#余下的 20%为测试集
test_dataset = dataset.drop(train_dataset.index)
train_stats = train_dataset.describe()
test_stats = test_dataset.describe()
#MPG 是训练模型要求的结果，不需要参与计算
train_stats.pop("MPG")
#对统计结果做转置
train_stats = train_stats.transpose()
#训练集和测试集都去掉 MPG 列，将其单独取出并作为标注
train_labels = train_dataset.pop('MPG')
test_labels = test_dataset.pop('MPG')
#定义一个数据规范化函数以简化操作
def norm(x):
    return (x - train_stats['mean']) / train_stats['std']
#训练集和测试集数据规范化
normed_train_data = norm(train_dataset)
```

```
normed_test_data = norm(test_dataset)
#输出规范化的训练集和测试集
print( " normed_train_data " )
print(normed_train_data)
print( " \n " )
print( " normed_test_data " )
print(normed_test_data)
```

（13）代码的运行结果如下。

```
normed_train_data
Cylinders    Displacement   Horsepower    Weight       Acceleration   Model
Year         USA            Europe        Japan
146          -0.869348      -1.009459     -0.784052    -1.025303      -0.379759
-0.516397    0.774676       -0.465148     -0.495225
282          -0.869348      -0.530218     -0.442811    -0.118796      0.624102
0.843910     0.774676       -0.465148     -0.495225
69           1.483887       1.482595      1.447140     1.736877       -0.738281
-1.060519    0.774676       -0.465148     -0.495225
378          -0.869348      -0.865687     -1.099044    -1.025303      -0.308055
1.660094     0.774676       -0.465148     -0.495225
331          -0.869348      -0.942365     -0.994047    -1.001603      0.875068
1.115971     -1.286751      -0.465148     2.012852
...          ...            ...           ...          ...            ...
...          ...            ...           ...
281          0.307270       0.044872      -0.521559    -0.000298      0.946772
0.843910     0.774676       -0.465148     -0.495225
229          1.483887       1.961837      1.972127     1.457223       -1.598734
0.299787     0.774676       -0.465148     -0.495225
150          -0.869348      -0.836932     -0.311564    -0.710099      -0.021237
-0.516397    -1.286751      -0.465148     2.012852
145          -0.869348      -1.076553     -1.151543    -1.169870      1.233589
-0.516397    -1.286751      -0.465148     2.012852
182          -0.869348      -0.846517     -0.495310    -0.623596      -0.021237
0.027726     -1.286751      2.143005      -0.495225

[314 rows x 9 columns]
normed_test_data
Cylinders    Displacement   Horsepower    Weight       Acceleration   Model
Year         USA            Europe        Japan
9            1.483887       1.865988      2.234620     1.018782       -2.530891
-1.604642    0.774676       -0.465148     -0.495225
25           1.483887       1.578444      2.890853     1.925289       -0.559020
-1.604642    0.774676       -0.465148     -0.495225
28           1.483887       1.041693      2.313368     2.063931       1.054328
-1.604642    0.774676       -0.465148     -0.495225
```

31	-0.869348	-0.789008	-0.259066	-0.903250	-0.559020
-1.332580	-1.286751	-0.465148	2.012852		
33	0.307270	0.351586	-0.127819	-0.422150	-0.917542
-1.332580	0.774676	-0.465148	-0.495225		
...	...	...	...	...	...
...	...	...	...		
369	-0.869348	-0.798593	-0.442811	-0.705359	0.875068
1.660094	0.774676	-0.465148	-0.495225		
375	-0.869348	-0.865687	-0.810302	-1.197124	-0.092942
1.660094	-1.286751	2.143005	-0.495225		
382	-0.869348	-0.836932	-0.915299	-0.883106	0.480693
1.660094	-1.286751	-0.465148	2.012852		
384	-0.869348	-0.999874	-0.994047	-1.214899	0.050467
1.660094	-1.286751	-0.465148	2.012852		
396	-0.869348	-0.721914	-0.679055	-0.432815	1.090181
1.660094	0.774676	-0.465148	-0.495225		

[78 rows x 9 columns]

（14）构建回归模型并显示模型详情，代码如下。

```
from tensorflow import keras
import pandas as pd
import ssl
from tensorflow.keras import layers
import tensorflow as tf
ssl._create_default_https_context = ssl._create_unverified_context
#定义下载地址
dataset_path=keras.utils.get_file("auto-mpg.data","http://archive.ics.uci.edu/ml/machine-learning-databases/auto-mpg/auto-mpg.data")
#规定列的名称
column_names = ['MPG','Cylinders','Displacement','Horsepower','Weight',
    'Acceleration', 'Model Year', 'Origin']
#读取数据
raw_dataset = pd.read_csv(dataset_path, names=column_names,
    na_values = "?", comment='\t', sep=" ", skipinitialspace=True)
#复制一份数据做后续操作
dataset = raw_dataset.copy()
#去掉无效数据
dataset = dataset.dropna()
#对 Origin 数据做 one-hot 编码，将数字转换成相应的产地
origin = dataset.pop('Origin')
dataset['USA'] = (origin == 1)*1.0
dataset['Europe'] = (origin == 2)*1.0
dataset['Japan'] = (origin == 3)*1.0
#通过 frac 参数随机分配 80%的数据，并将其作为训练集
```

```
#random_state 是随机数的种子,使用固定的值保证每次构建数据集的数据一致
train_dataset = dataset.sample(frac=0.8, random_state=0)
#余下的 20%为测试集
test_dataset = dataset.drop(train_dataset.index)
train_stats = train_dataset.describe()
test_stats = test_dataset.describe()
#MPG 是训练模型要求的结果,不需要参与计算
train_stats.pop("MPG")
#对统计结果做转置
train_stats = train_stats.transpose()
#训练集和测试集都去掉 MPG 列,将其单独取出并作为标注
train_labels = train_dataset.pop('MPG')
test_labels = test_dataset.pop('MPG')
#定义一个数据规范化函数以简化操作
def norm(x):
    return (x - train_stats['mean']) / train_stats['std']
#训练集和测试集数据规范化
normed_train_data = norm(train_dataset)
normed_test_data = norm(test_dataset)
#构建回归模型,模型为顺序模型,使用的激活算法为 relu
def build_model():
    model = keras.Sequential([
        layers.Dense(64, activation='relu', input_shape=[len(train_dataset.keys())]),
        layers.Dense(64, activation='relu'),
        layers.Dense(1)
    ])
    optimizer = tf.keras.optimizers.RMSprop(0.001)
    model.compile(loss='mse', optimizer=optimizer, metrics=['acc'])
    return model
#构建模型
model = build_model()
#显示模型详情
model.summary()
```

(15) 代码的运行结果如下。

```
Model: "sequential"
```

Layer (type)	Output Shape	Param #
dense (Dense)	(None, 64)	640
dense_1 (Dense)	(None, 64)	4160
dense_2 (Dense)	(None, 1)	65

```
=================================================================
Total params: 4,865
Trainable params: 4,865
Non-trainable params: 0
_____
```

（16）使用训练集对模型进行训练，本例中训练1000次，代码如下。

```
from tensorflow import keras
import pandas as pd
import ssl
from tensorflow.keras import layers
import tensorflow as tf
ssl._create_default_https_context = ssl._create_unverified_context
#定义下载地址
dataset_path=keras.utils.get_file("auto-mpg.data","http://archive.ics.uci.edu/ml/machine-learning-databases/auto-mpg/auto-mpg.data")
#规定列的名称
column_names = ['MPG','Cylinders','Displacement','Horsepower','Weight',
    'Acceleration', 'Model Year', 'Origin']
#读取数据
raw_dataset = pd.read_csv(dataset_path, names=column_names,
    na_values = "?", comment='\t', sep=" ", skipinitialspace=True)
#复制一份数据做后续操作
dataset = raw_dataset.copy()
#去掉无效数据
dataset = dataset.dropna()
#对 Origin 数据做 one-hot 编码，将数字转换成相应的产地
origin = dataset.pop('Origin')
dataset['USA'] = (origin == 1)*1.0
dataset['Europe'] = (origin == 2)*1.0
dataset['Japan'] = (origin == 3)*1.0
#通过 frac 参数随机分配 80%的数据，并将其作为训练集
#random_state 是随机数的种子，使用固定的值保证每次构建数据集的数据一致
train_dataset = dataset.sample(frac=0.8, random_state=0)
#余下的 20%为测试集
test_dataset = dataset.drop(train_dataset.index)
train_stats = train_dataset.describe()
test_stats = test_dataset.describe()
#MPG 是训练模型要求的结果，不需要参与计算
train_stats.pop("MPG")
#对统计结果做转置
train_stats = train_stats.transpose()
#训练集和测试集都去掉 MPG 列，将其单独取出并作为标注
train_labels = train_dataset.pop('MPG')
```

```python
test_labels = test_dataset.pop('MPG')
#定义一个数据规范化函数以简化操作
def norm(x):
    return (x - train_stats['mean']) / train_stats['std']
#训练集和测试集数据规范化
normed_train_data = norm(train_dataset)
normed_test_data = norm(test_dataset)
#构建回归模型,模型为顺序模型,使用的激活算法为relu
def build_model():
    model = keras.Sequential([
        layers.Dense(64, activation='relu', input_shape=[len(train_dataset.keys())]),
        layers.Dense(64, activation='relu'),
        layers.Dense(1)
    ])
    optimizer = tf.keras.optimizers.RMSprop(0.001)
    model.compile(loss='mse', optimizer=optimizer, metrics=['acc'])
    return model
model = build_model()
#制订训练计划并执行
EPOCHS = 1000
history = model.fit(normed_train_data, train_labels,
    epochs=EPOCHS, validation_split=0.2, verbose=0)
#使用测试集预测数据
test_result = model.predict(normed_test_data)
#显示预测结果
print('==================\ntest_result:', test_result)
```

(17) 代码的运行结果如下。

```
test_result: [[15.674531]
 [10.523621]
 ...
 [29.77305 ]]
```

(18) 上述结果是数字化的预测结果,下面显示训练数据的图像,即平均绝对误差(MAE)和均方误差(MSE),代码如下。

```
from tensorflow import keras
import pandas as pd
import ssl
from tensorflow.keras import layers
import tensorflow as tf
import matplotlib.pyplot as plt
ssl._create_default_https_context = ssl._create_unverified_context
#定义下载地址
dataset_path=keras.utils.get_file("auto-mpg.data","http://archive.ics.uci.edu/ml/machine-learning-databases/auto-mpg/auto-mpg.data")
```

```python
#规定列的名称
column_names = ['MPG','Cylinders','Displacement','Horsepower','Weight',
    'Acceleration', 'Model Year', 'Origin']
#读取数据
raw_dataset = pd.read_csv(dataset_path, names=column_names,
    na_values = "?", comment='\t', sep=" ", skipinitialspace=True)
#复制一份数据做后续操作
dataset = raw_dataset.copy()
#去掉无效数据
dataset = dataset.dropna()
#对 Origin 数据做 one-hot 编码，将数字转换成相应的产地
origin = dataset.pop('Origin')
dataset['USA'] = (origin == 1)*1.0
dataset['Europe'] = (origin == 2)*1.0
dataset['Japan'] = (origin == 3)*1.0
#通过 frac 参数随机分配 80%的数据，并将其作为训练集
#random_state 是随机数的种子，使用固定的值保证每次构建数据集的数据一致
train_dataset = dataset.sample(frac=0.8, random_state=0)
#余下的 20%为测试集
test_dataset = dataset.drop(train_dataset.index)
train_stats = train_dataset.describe()
test_stats = test_dataset.describe()
#MPG 是训练模型要求的结果，不需要参与计算
train_stats.pop("MPG")
#对统计结果做转置
train_stats = train_stats.transpose()
#训练集和测试集都去掉 MPG 列，将其单独取出并作为标注
train_labels = train_dataset.pop('MPG')
test_labels = test_dataset.pop('MPG')
#定义一个数据规范化函数以简化操作
def norm(x):
    return (x - train_stats['mean']) / train_stats['std']
#训练集和测试集数据规范化
normed_train_data = norm(train_dataset)
normed_test_data = norm(test_dataset)
#构建回归模型，模型为顺序模型，使用的激活算法为 relu
def build_model():
    model = keras.Sequential([
        layers.Dense(64, activation='relu', input_shape=[len(train_dataset.keys())]),
        layers.Dense(64, activation='relu'),
        layers.Dense(1)
    ])
```

```python
    optimizer = tf.keras.optimizers.RMSprop(0.001)
    model.compile(loss='mse', optimizer=optimizer, metrics=['mae', 'mse'])
    return model
model = build_model()
#制订训练计划并执行
EPOCHS = 1000
history = model.fit(normed_train_data, train_labels,
    epochs=EPOCHS, validation_split=0.2, verbose=0)
#使用测试集预测数据
test_result = model.predict(normed_test_data)
#定义图像输出形式
def plot_history(history):
    hist = pd.DataFrame(history.history)
    hist['epoch'] = history.epoch
    plt.figure('MAE --- MSE', figsize=(8, 4))
    plt.subplot(1, 2, 1)
    plt.xlabel('epoch')
    plt.ylabel('Mean Absolute Error (MPG)')
    plt.plot(
        hist['epoch'], hist['mae'],
        label='train error')
    plt.plot(
        hist['epoch'], hist['val_mae'],
        label='val error')
    plt.ylim([0, 5])
    plt.legend()
    plt.subplot(1, 2, 2)
    plt.xlabel('epoch')
    plt.ylabel('Mean Square Error ($MPG^2$)')
    plt.plot(
        hist['epoch'], hist['mse'],
        label='train error')
    plt.plot(
        hist['epoch'], hist['val_mse'],
        label='val error')
    plt.ylim([0, 20])
    plt.legend()
    plt.show()
#显示训练数据的图像
plot_history(history)
```

（19）运行代码，得到 MAE 和 MSE 图如图 13-5 所示。

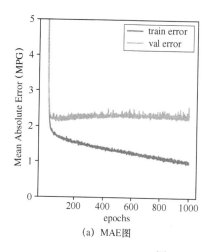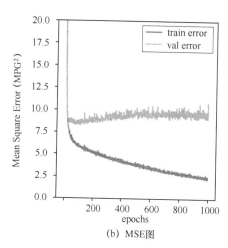

图 13-5　MAE 图和 MSE 图

从图中可以看出，虽然随着迭代次数的增加，训练错误率逐渐降低，但在达到一定次数之后，训练错误率逐渐稳定，此时的训练次数明显是无效的。下面对这种情况进行优化。

（20）TensorFlow 2.0 提供了 EarlyStopping 回调函数来对上述情况进行优化。

```
from tensorflow import keras
import pandas as pd
import ssl
from tensorflow.keras import layers
import tensorflow as tf
import matplotlib.pyplot as plt
ssl._create_default_https_context = ssl._create_unverified_context
#定义下载地址
dataset_path=keras.utils.get_file("auto-mpg.data","http://archive.ics.uci.edu/ml/machine-learning-databases/auto-mpg/auto-mpg.data")
#规定列的名称
column_names = ['MPG','Cylinders','Displacement','Horsepower','Weight',
    'Acceleration', 'Model Year', 'Origin']
#读取数据
raw_dataset = pd.read_csv(dataset_path, names=column_names,
    na_values = "?", comment='\t', sep=" ", skipinitialspace=True)
#复制一份数据做后续操作
dataset = raw_dataset.copy()
#去掉无效数据
dataset = dataset.dropna()
#对 Origin 数据做 one-hot 编码，将数字转换成相应的产地
origin = dataset.pop('Origin')
dataset['USA'] = (origin == 1)*1.0
dataset['Europe'] = (origin == 2)*1.0
dataset['Japan'] = (origin == 3)*1.0
#通过 frac 参数随机分配 80%的数据，并将其作为训练集
```

```python
#random_state 是随机数的种子，使用固定的值保证每次构建数据集的数据一致
train_dataset = dataset.sample(frac=0.8, random_state=0)
#余下的 20%为测试集
test_dataset = dataset.drop(train_dataset.index)
train_stats = train_dataset.describe()
test_stats = test_dataset.describe()
#MPG 是训练模型要求的结果，不需要参与计算
train_stats.pop("MPG")
#对统计结果做转置
train_stats = train_stats.transpose()
#训练集和测试集都去掉 MPG 列，将其单独取出并作为标注
train_labels = train_dataset.pop('MPG')
test_labels = test_dataset.pop('MPG')
#定义一个数据规范化函数以简化操作
def norm(x):
    return (x - train_stats['mean']) / train_stats['std']
#训练集和测试集数据规范化
normed_train_data = norm(train_dataset)
normed_test_data = norm(test_dataset)
#构建回归模型，模型为顺序模型，使用的激活算法为 relu
def build_model():
    model = keras.Sequential([
        layers.Dense(64, activation='relu', input_shape=[len(train_dataset.keys())]),
        layers.Dense(64, activation='relu'),
        layers.Dense(1)
    ])
    optimizer = tf.keras.optimizers.RMSprop(0.001)
    model.compile(loss='mse', optimizer=optimizer, metrics=['mae', 'mse'])
    return model
model = build_model()
EPOCHS = 1000
#设置 EarlyStopping 回调函数，监控 val_loss 指标
#当该指标在 10 次迭代中均不变化后退出
early_stop = keras.callbacks.EarlyStopping(monitor='val_loss', patience=10)
#对加入了回调函数的模型进行训练
history = model.fit(normed_train_data, train_labels, epochs=EPOCHS,
    validation_split = 0.2, verbose=0, callbacks=[early_stop])
#使用测试集预测数据
test_result = model.predict(normed_test_data)
#定义图像输出形式
def plot_history(history):
    hist = pd.DataFrame(history.history)
    hist['epoch'] = history.epoch
    plt.figure('MAE --- MSE', figsize=(8, 4))
```

```python
    plt.subplot(1, 2, 1)
    plt.xlabel('epoch')
    plt.ylabel('Mean Absolute Error(MPG)')
    plt.plot(
        hist['epoch'], hist['mae'],
        label='train error')
    plt.plot(
        hist['epoch'], hist['val_mae'],
        label='val error')
    plt.ylim([0, 5])
    plt.legend()
    plt.subplot(1, 2, 2)
    plt.xlabel('epoch')
    plt.ylabel('Mean Square Error(MPG^2)')
    plt.plot(
        hist['epoch'], hist['mse'],
        label='train error')
    plt.plot(
        hist['epoch'], hist['val_mse'],
        label='val error')
    plt.ylim([0, 20])
    plt.legend()
    plt.show()
#显示图像
plot_history(history)
```

（21）运行代码，得到加入 EarlyStopping 回调函数的 MAE 和 MSE 图如图 13-6 所示。

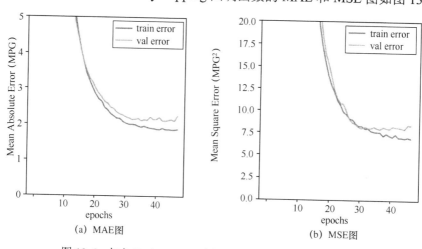

图 13-6　加入 EarlyStopping 回调函数的 MAE 图和 MSE 图

加入 EarlyStopping 回调函数能够减少无效的训练次数，具体效果可以从图 13-5 和图 13-6 的对比中看出。

（22）在模型训练完成并基本达到要求后，可以对其进行预测。下面的代码对预测的结果进行了对比。

```python
from tensorflow import keras
import pandas as pd
import ssl
from tensorflow.keras import layers
import tensorflow as tf
import matplotlib.pyplot as plt
ssl._create_default_https_context = ssl._create_unverified_context
#定义下载地址
dataset_path=keras.utils.get_file("auto-mpg.data","http://archive.ics.uci.edu/ml/machine-learning-databases/auto-mpg/auto-mpg.data")
#规定列的名称
column_names = ['MPG','Cylinders','Displacement','Horsepower','Weight',
    'Acceleration', 'Model Year', 'Origin']
#读取数据
raw_dataset = pd.read_csv(dataset_path, names=column_names,
    na_values = "?", comment='\t', sep=" ", skipinitialspace=True)
#复制一份数据做后续操作
dataset = raw_dataset.copy()
#去掉无效数据
dataset = dataset.dropna()
#对 Origin 数据做 one-hot 编码，将数字转换成相应的产地
origin = dataset.pop('Origin')
dataset['USA'] = (origin == 1)*1.0
dataset['Europe'] = (origin == 2)*1.0
dataset['Japan'] = (origin == 3)*1.0
#通过 frac 参数随机分配 80%的数据，并将其作为训练集
#random_state 是随机数的种子，使用固定的值保证每次构建数据集的数据一致
train_dataset = dataset.sample(frac=0.8, random_state=0)
#余下的 20%为测试集
test_dataset = dataset.drop(train_dataset.index)
train_stats = train_dataset.describe()
test_stats = test_dataset.describe()
#MPG 是训练模型要求的结果，不需要参与到计算中
train_stats.pop("MPG")
#对统计结果做转置
train_stats = train_stats.transpose()
#训练集和测试集都去掉 MPG 列，将其单独取出并作为标注
train_labels = train_dataset.pop('MPG')
test_labels = test_dataset.pop('MPG')
#定义一个数据规范化函数以简化操作
def norm(x):
```

```python
    return (x - train_stats['mean']) / train_stats['std']
#训练集和测试集数据规范化
normed_train_data = norm(train_dataset)
normed_test_data = norm(test_dataset)
#构建回归模型，模型为顺序模型，使用的激活算法为relu
def build_model():
    model = keras.Sequential([
        layers.Dense(64, activation='relu', input_shape=[len(train_dataset.keys())]),
        layers.Dense(64, activation='relu'),
        layers.Dense(1)
    ])
    optimizer = tf.keras.optimizers.RMSprop(0.001)
    model.compile(loss='mse', optimizer=optimizer, metrics=['mae', 'mse'])
    return model
model = build_model()
EPOCHS = 1000
#设置EarlyStopping回调函数，监控val_loss指标
#当该指标在10次迭代中均不变化后退出
early_stop = keras.callbacks.EarlyStopping(monitor='val_loss', patience=10)
#对加入了回调函数的模型进行训练
history = model.fit(normed_train_data, train_labels, epochs=EPOCHS,
    validation_split = 0.2, verbose=0, callbacks=[early_stop])
#使用测试集数据预测模型
test_predictions = model.predict(normed_test_data).flatten()
#绘制标注相同的对比图和误差分布直方图
plt.figure('Prediction & TrueValues  --- Error', figsize=(8, 4))
plt.subplot(1, 2, 1)
plt.scatter(test_labels, test_predictions)
plt.xlabel('true values(MPG)')
plt.ylabel('predictions(MPG)')
plt.axis('equal')
plt.axis('square')
plt.xlim([0, plt.xlim()[1]])
plt.ylim([0, plt.ylim()[1]])
_ = plt.plot([-100, 100], [-100, 100])
error = test_predictions - test_labels
plt.subplot(1, 2, 2)
plt.hist(error, bins=25)
plt.xlabel(" prediction error(MPG) ")
_ = plt.ylabel(" count ")
plt.show()
```

（23）运行代码，得到对比图和误差分布直方图如图13-7所示。

从图13-7中可以看出，预测的结果与标注的点基本一致。

第 13 章 TensorFlow 回归

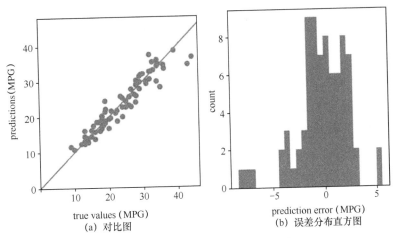

(a) 对比图　　(b) 误差分布直方图

图 13-7　对比图和误差分布直方图

反侵权盗版声明

电子工业出版社依法对本作品享有专有出版权。任何未经权利人书面许可，复制、销售或通过信息网络传播本作品的行为；歪曲、篡改、剽窃本作品的行为，均违反《中华人民共和国著作权法》，其行为人应承担相应的民事责任和行政责任，构成犯罪的，将被依法追究刑事责任。

为了维护市场秩序，保护权利人的合法权益，我社将依法查处和打击侵权盗版的单位和个人。欢迎社会各界人士积极举报侵权盗版行为，本社将奖励举报有功人员，并保证举报人的信息不被泄露。

举报电话：（010）88254396；（010）88258888
传　　真：（010）88254397
E-mail：　dbqq@phei.com.cn
通信地址：北京市万寿路173信箱
　　　　　电子工业出版社总编办公室
邮　　编：100036